Learn to See the Invisible

Most improvement consultants say improvement efforts must be led by the CEO, and that is certainly ideal. But the actual reality is most CEOs do not actively drive/guide improvement. They want it to happen, but they focus most of their energy on other issues. According to surveys from Gallup and others, the number one reason people say, "I am not engaged" is due to the behaviors of their direct boss! Those leaders (in the middle of an organization) have a tremendous amount of leverage; first- and second-line leaders directly touch 80% of the people in their organization. They have a tremendous amount of influence and more power than they might realize. This book focuses on that demographic.

This book describes four key foundations and 25 different actions leaders can practice to become more effective in training their eyes to see things tomorrow that are currently invisible. It helps leaders and managers to become better observers of their current reality by practicing getting better at getting better. The goal is to get better in the way we lead, the way our team performs, and the results we accomplish. If done in the right way, visually posting your improvement targets is the key to driving more personal growth, as well as more cross-functional collaboration and cooperation in your work activities. The most unique aspect of this book is that it suggests leaders use visual tools.

Visual Leadership is the fourth foundational element the author encourages people to practice. The primary purpose of visual performance metrics isn't to make sure things are working well in your department. It's to create a thinking environment that makes it easier for multiple departments, teams, and groups to work together. It is relatively easy to come up with performance metrics for your team, but what about metrics that help "us" to work more effectively together? Good visual reporting practices create "information democracy." They eliminate filters that obscure knowledge between layers of management and between departments. They help to create an environment that is much more robust and open, making it easier to be in touch with the "actual reality." And perhaps the most exciting of all, visual tools can help an individual learn to lead more effectively. The power of using visuals in this way is underutilized in most organizations.

Whatever new behaviors a leader is trying to accomplish, visual reporting can facilitate progress and ensure accountability in developing those new habits.

"*Learn to See the Invisible* is not just another leadership book. It's a book about how to produce excellence in a profoundly challenging world by respecting, developing and utilizing the intelligence of all employees. Michael's deep immersion in lean leadership, and awareness of Toyota practices, has given him invaluable expertise that he now generously shares with you. Gleaned from acutely sensitive observations of work – and the people who do it – Bremer's wisdom shines through in this valuable resource for anyone who wants to be a better leader."

Amy C. Edmondson, *Novartis Professor of Leadership, Harvard Business School; Author,* Right Kind of Wrong: The Science of Failing Well (Atria 2023)

"Michael knows his way around the Gemba better than anyone I know. And he knows what he's looking for. He's visited with our teams more times than I can count, and he always comes back with great insights. His knowledge and experience show in this book.

As I read it, I repeatedly thought, 'These are the things managers need to be doing. These are the things I would tell them to do. Pick a few of the suggestions and begin improving your leadership!"

Gary Peterson, *EVP Supply Chain and Production, O.C. Tanner Company*

"We used Michael's book, *Learn to See the Invisible*, as a basis of a 12-week experiment, designed to combine learning and action with a handful of factory leaders, from a variety of positions within our organization. We walked with them through a journey of discovery about the actions they could take to be the leader they wanted to be and help their teams be successful. Michael's book provided real-world case studies, practical tools, and thought-provoking questions for cohort. It was amazing the difference we saw over those 3 months! The program participants were taken aback when they reflected on how much had changed during that time with one of them stating, 'Before we started this activity, I had several areas that were struggling to meet their numbers. As I applied what I was learning from the book, I could see us improving. Those same areas now have higher team member engagement and are consistently meeting their performance targets with less directive instruction from me!'

We believe that the content found in this book, when applied, will support any leader in delivering consistently better results. So the question is, 'What are **you** waiting for?'"

Cindy Hinds, *Global Director of Enterprise Excellence & Jim Gunville, Director of Operations for North America Water Treatment, A. O. Smith Corporation*

"In his book, *Learn to See the Invisible*, Michael brings a lifetime of lessons from working with high-performing teams and helping others get there. The wisdom shared is concise and filled with practical ideas that can be tried tomorrow. Read and apply these lessons and your work life will be enriched, and the teams you lead will thank you for making important changes to your leadership efforts."

Richard Sheridan, *CEO and Chief Storyteller at Menlo Innovations; Speaker, Author of* Chief Joy Officer *and* Joy, Inc.

"In various sectors like tech, bio-tech, and heavy industry, everything from regulations to market access is uniform within each sector. However, outcomes vary greatly. The few best dominate their field by huge margins, top performers excel in all aspects by consistently implementing Michael Bremer's ideas. This involves leaders actively participating in the work process, engaging with the workforce, and fostering widespread capability development for continuous problem-solving from diverse perspectives."

Steve Spear *DBA MS MS, author of* The High Velocity Edge *and* Wiring the Winning Organization *and MIT Sloan School of Management, Senior Lecturer*

"Michael is the go-to person for best practices in leadership and respect for people. His book, *Learn to See the Invisible,* is not your typical leadership book. Michael shares how to 'train our eyes and ears to see things tomorrow that lie invisible today.' In it, Michael shares his experience and research in a thought-provoking style and challenges readers to strive for excellence. He provides real-life actionable insights that resonate with anyone on their journey to excellence. Readers will learn to operate at a championship level of leadership performance. Michael's foundation of Reflection, Unifying Purpose, Building Relationships, and Visual Leadership empowers leaders to grow, transform, and improve. Michael's book challenges us to be a better leader."

Kim Humphrey, *President/CEO Association for Manufacturing Excellence*

Learn to See the Invisible

How to Unlock Your Potential as a Leader

Michael Bremer

A PRODUCTIVITY PRESS BOOK

First published 2025
by Routledge
605 Third Avenue, New York, NY 10158

and by Routledge
4 Park Square, Milton Park, Abingdon, Oxon, OX14 4RN

Routledge is an imprint of the Taylor & Francis Group, an informa business

ISBN: 9781032800714 (hbk)
ISBN: 9781032800707 (pbk)
ISBN: 9781003495284 (ebk)

DOI: 10.4324/9781003495284

Typeset in Garamond
by Deanta Global Publishing Services, Chennai, India

Contents

Preface

Why did I write this book? I worked in the field of organizational performance improvement for more than 40 years. When I first began this type of work, I did not really know what highly effective organizational improvement practices looked like. Forty years later, I finally do know something.

Today, when I look at how most organizations go about the business of improvement, it's quite clear that most organizational senior leaders also don't know a whole lot about highly effective and holistic improvement practices. Interestingly, I've had opportunities to observe and interact with a few leaders who do get it. Many of those people started from the middle of their organization when they embarked on their journey. They were not the CEOs.

What I hope to accomplish through these few pages is to challenge the reader, who is also most likely not the CEO. What are some practical steps anyone can take to become a better leader? The four foundational steps shared in this book come from personal experience and mirror the steps followed by those few leaders who get it; they have done an outstanding job of creating an environment where their organization can flourish.

The overwhelming feedback on my last book, *How to Do a Gemba Walk*, was its practicality. It lays out the basic steps someone needs to do to be effective with Gemba Walks inside their organization. I hope to accomplish something similar with *Learn to See the Invisible* – practical steps any leader can practice to get better at getting better.

This book's closing sentence states: "I genuinely hope you embark on this journey. Experiment, learn, grow, and share with others. The world needs better leadership. Please provide it."

There is also a group of individuals who read early drafts of this book, shared their stories to include in the text and challenged my thinking.

Figures 1.4, 2.1, 2.2, 3.3, 4.1, 4.4, 4.8, and 4.9 were enhanced by Michael Bremer using DALL-E.

Acknowledgments

This book would not have come about without a whole lot of help from other people. First of all, I must thank my wife, Lynn Sieben, of 45+ years for her support and encouragement. She was an upgrade in my life. My life is so much better than it might have otherwise been without her as a partner by my side. Who knew that meeting a woman on a Chicago #22 bus could result in such an outcome?

I must also thank the many organizations that let me visit their site multiple times to try to gain a deeper understanding of how they truly did become an organization that was highly effective at improving. Those companies included: O.C. Tanner, Autoliv, Cogent, Menlo Innovations, and the many award recipients and applicants from my AME Excellence Award activities.

There is also a group of individuals who read early drafts of this book, shared their stories and challenged my thinking. They include: Gary Peterson, Tom Hartman, Richard Sheridan, Ron Harper, Jim Garrick, Mark Preston, Dan McDonnell, Art Byrne, Anu George, George Koenigsacker, Kevin Meyer, Cindy Hinds (who actually let some of the emerging leaders at A.O. Smith test practice the four foundations), Allan Coletta, Cheryl Jekiel, Jim Glover, Victor Caune, Marc Kurzic, Sheila Carroll (a life-long friend), Coby Stoller, Erin Riley, Steven Spear, Ralf VanSosen, James Hammond, Will Wiesner, Brad Jeavons, David Bovis, and Kim Humphrey (CEO for AME).

About the Author

Michael Bremer semi-retired from business in 2018. He participated in a variety of experiences over the years, including Director of Productivity Improvement and later Director of Information Systems for Beatrice Foods; past Chief Financial Officer and Board Member for the Association of Manufacturing Excellence (AME); and President of the Cumberland Group in Chicago for 28 years (global consulting company). Mr. Bremer also served as adjunct faculty for the University of Chicago's Graham School for a 15-year period and also served as a senior mentor at a new business start-up incubator focused on manufacturing (mHub Chicago).

Awards:

- 2016 Shingo Prize Research Award for *How to Do a Gemba Walk*
- 2019 recipient of the Mac McCulloch Life Time Achievement Award from AME

Written five books:

1. *Six Sigma Black Belt Handbook* – 2005 McGraw Hill Technical Publishing
2. *Six Sigma Financial Tracking & Reporting* – 2007 McGraw Hill Technical Publishing

3. *Escape the Improvement Trap* – 2010 Taylor & Francis
4. *How to Do a Gemba Walk* – 2016 Self-Published
5. *Cómo Hacer un Gemba Walk* – 2020 Profit Editorial (Spanish version)
6. *Learn to See the Invisible: How to Unlock Your Potential as a Leader* – 2024 CRC Press

Some of his current and recent activities:

■ Serves as a volunteer V.P. for the Association of Manufacturing Excellence Award activities
   a. Manages a network of 50+ volunteer assessors
   b. Led a site visit team to Malaysia 2022, China 2023
■ Moderates a bi-weekly Global Issues discussion group
■ In first year of retirement managed to travel 120,000 miles, which was kind of interesting
■ Hobbies: running, skiing, golf
■ Married to my best friend – Lynn (45 years!!!)

Moved from Chicago to San Francisco area to be closer to grandchildren. My wife and I are currently very much enjoying life.

Feel free to connect with me on LinkedIn, Twitter @michaelbremer; you will find many of the worksheets shown in this book at my website http://michaelbremer.net.

Chapter 1

Introduction

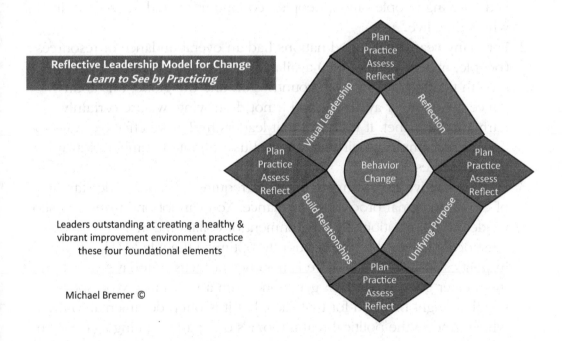

Reflective Leadership Model for Change
Learn to See by Practicing

Leaders outstanding at creating a healthy &
vibrant improvement environment practice
these four foundational elements

Michael Bremer ©

Purpose of This Book

The primary purpose of this book is to help leaders become more effective and inclusive, and to train their eyes/ears to see things tomorrow that
lay invisible today. It's also intended to increase their ability to inspire team
members and peers to:

DOI: 10.4324/9781003495284-1

- Grow people's capabilities to do things they never thought possible and experience more joy in their work.
- Tap into that optional treasure where people become passionate and inspired by their work, finding more meaning and fulfillment.

The intent is to provide a few practical steps anyone can practice discovering the treasure and to help leaders at multiple levels of an organization address three critical challenges:

1. The immense underutilization of talents inside most companies. People could do much more and live more fulfilling lives if their organization (services, manufacturing, governmental entity, or nonprofit) was run more holistically and effectively. Inadequate and mediocre leadership is destroying people's lives, people's communities, and the society in which they live.
2. For many years, developed nations had an overabundance of resources (people, materials, land, etc.) available for their use. With industrial growth taking place in more countries around the globe, this abundance is no longer available, and if not destroying, we are certainly harming the planet. It is critical that leaders find more effective ways to operate, accomplish their mission, and use a limited number of available resources.
3. Highly effective improvement practices require a deeper understanding of cross-functional process performance. You cannot optimize each silo inside an organization. An environment of collaboration and harmony must occur across the organization to optimize the whole. This is true in politics and business. There is insufficient focus in today's world on the greater good. Too much gets done from a very narrow perspective that might be good for that slice, but it is often detrimental to the whole. And in the political realm too, it's one party fighting against the other party rather than working together to improve the situation for the whole. We must change how we think and behave relative to working with others for highly effective leadership practices.

Organizational leaders, at any level, can help the people around them to grow. They can help their teammates, peers, and even their boss see the invisible. Business and governmental leadership can become a more robust and positive force for good if we are bold in changing how we currently operate.

I intend this book to be useful for anyone in a leadership position to grow their capabilities. Mid-level leaders far outnumber the number of CEOs; they can make the world a better place.

The Reflective Leadership Model for Change

In this book, we discuss four foundational levers for changing the way you lead and improving in a more effective way (Figure 1.1).

- **Step 1: Reflection:** Get a handle on the "actual reality" of how you currently lead. In this initial step, leaders embark on a journey of self-discovery and self-awareness. They engage in introspection and seek feedback to gain a clear understanding of their existing leadership style, strengths, weaknesses, and opportunities for improvement. The reflective process helps leaders to identify their starting point and sets the foundation for meaningful change.
- **Step 2: Unifying Purpose:** Create a meaningful and real purpose to guide behavior change. The purpose serves as a guiding light to inspire and motivate both the leader and the team. It should resonate with the values and aspirations of the person(s) making the change. Use it to target a desirable future state. Measure your progress toward fulfilling the purpose.

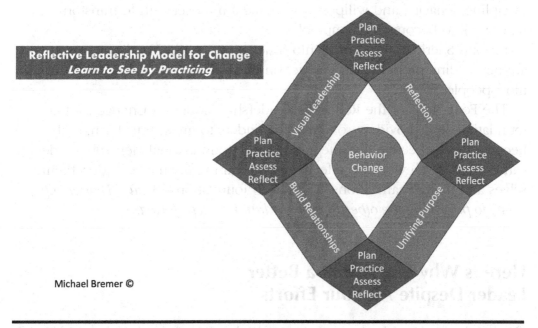

Figure 1.1   Reflective leadership model for change

- **Step 3: Build Relationships:** Elevate people and mindset. Leaders build stronger relationships with their team members, peers, and stakeholders in this step. They cultivate an environment of trust, respect, and open communication. Leaders empower and elevate those around them by actively listening, supporting, and recognizing individual strengths. Simultaneously, leaders foster a growth mindset within themselves and their teams, promoting continuous learning, adaptability, and a shared commitment to personal and collective development.
- **Step 4: Visual Leadership:** Create a thinking environment. A critical step involves implementing a visual framework that causes us to think and adjust the behaviors we strive to change. Visuals should obviously measure progress; they should also stimulate our brains and keep us accountable for desired changes. Great visuals inspire improvement rather than just compliance. Regular communication and discussion around these visual representations reinforce the purpose and let us celebrate successes. Through further reflection, we identify improvement areas and drive desirable change forward.

It does not matter if we look at team leaders, department leaders, or senior executives; most people think they already do a pretty good job of leading and communicating and that if they improve a little, that should move the organization to where it needs to go. Leaders do not typically possess the discipline, tenacity, and willpower to make a mindset shift to transform. It's hard work to become a champion!

Richard Sheridan, CEO of Menlo Innovations, said, "You know when you are not leading properly. Do you act on those moments of reflection? I think most people do not."[1]

The Four Steps of the Reflective Leadership Model for Change form a foundation, and provide a roadmap for leaders to grow, transform, and become more effective. By integrating these foundational elements, leaders can create a framework for effective and meaningful change within themselves and their organizations. The fourth foundation – *Visual Leadership* – *is key to propelling the other three foundational steps forward.*

Here Is Why You Aren't a Better Leader Despite All Your Efforts

Most of the leaders I have met want to do the right thing, but they are busy and need more time to do everything that needs to be done. A reasonable definition of stress if you think about it. But what if some of this stress is unintentionally self-induced? What if we believe we are doing the right thing, but actually, we're not?

Recently, I observed a highly competent leadership team. They did many things exceptionally well. I observed the leadership team's Gemba Walk. Their stated purpose for the walk was *"To show people we care and to identify any safety problems"* [emphasis added] – a reasonable and somewhat noble intention. The right people were doing the walk, but the walk they were doing wasn't what they *stated or* should have been doing. During their walk, they did not speak to a single operator. The local lead of the operation joined in the walk, but there was no communication with that person about why the walk was being done that day or any sharing of information from previous walks. There was also no feedback loop to gauge whether the leadership team accomplished its intended purpose using this particular improvement practice – Gemba Walks.

The leadership team assumed that because they focused on "safety," it would be evident to their work associates that they cared. But that wasn't the impression they were imparting. Instead, their walk looked like a detailed inspection: Was all safety equipment in the right place, was the lockout board up-to-date, etc.? The plant's general manager, the VP of operations, and the other leaders who went along for the walk were doing work that their associates should have instead been doing.

It should not be the leader's responsibility to find these issues. Their primary responsibility was to establish an environment where individuals doing the work should take full accountability and proactively address any problems. Leaders could then focus more on the processes of maintaining a completely safe work environment. Conversations with their associates would highlight inhibitors that make work difficult and positively support their growth and development.

Our intentions as leaders are typically good, but our habits, assumptions, and beliefs can often impede a proper understanding of reality. Positively changing how we lead will require reflection on the way we lead today, and making sure the purpose for how we behave/operate today as a leader is the appropriate purpose.

How Aware Are We of What We Do Not Know?

How likely are there essential things happening in your organization, division, department, or even with your team that you do not currently see? Would you believe that is possible?

Most leaders I've met over the years might not say it, but they would think, "That is impossible! I'm pretty sure I don't see it all, but I'm sure I have a handle on what is important."

Unfortunately, it's an actual reality. For various reasons, leaders don't see behaviors that inhibit the team, the department, and the organization's ability to excel. There is a lot of baggage that impedes their ability to see the actual reality. It does not matter if we talk about leaders in the United States, Europe, China, Australia, etc. The cultural reasons may differ, but the result is the same. Leaders have the potential to lead more effectively, but their habits, beliefs, and assumptions inhibit their ability to realize their full potential. And some of what they don't see is very important.

What are some of the everyday things that cause leaders to be out of touch?

- People don't want to make their boss look bad to higher-level authorities.
- In the Western World, especially in the United States, people feel solving problems is their responsibility to "figure it out." When associates are asked, "How is it going?" they respond, "It's OK."
- Providing good feedback to team members is complex, and many people like to avoid conflict because it makes them uncomfortable.
- Our habits can inhibit our learning. Habits are wired deeply in our brains; they don't require much thought, and we operate automatically.
- In some societies, there is a high deference to authority (or higher-level leaders), so cultural norms make it difficult to see what is happening.
- People learn to avoid pointing out problems or sharing their thoughts on improvement when nothing happens with their ideas.
- A few leaders don't want their subordinates telling them something is wrong, that they need to lead differently, or that they made a mistake. This type of leader gets surrounded by "YES! People." Unfortunately, some people who operate this way still become CEOs or high-level politicians, and good things rarely follow.

The list could go on, but I'm sure you get the idea. It begs the question:

What is the nature of the work environment for my team? As a leader, do I treat my people with the respect they deserve? Am I accomplishing my intended purpose? Do I actively seek to find practical ways to lead my team more effectively? Do I challenge myself to learn to see things that are currently invisible?

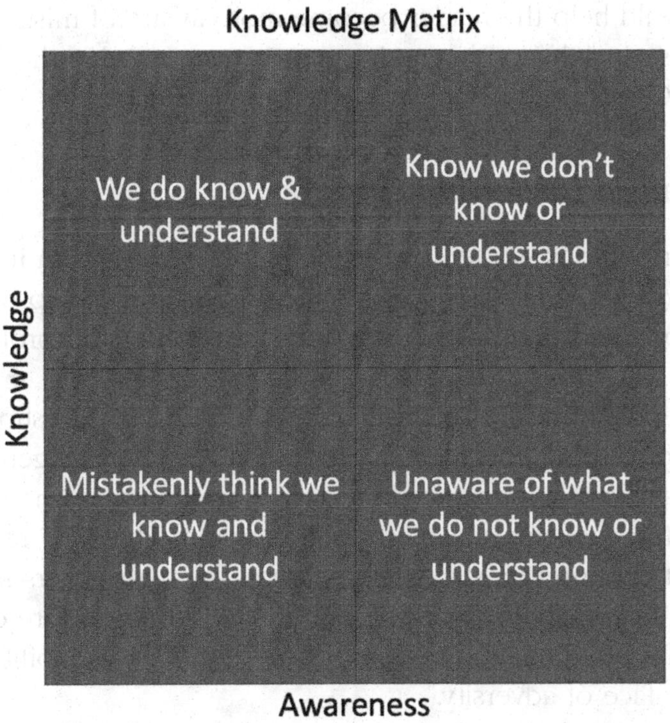

Knowledge Matrix

We do know & understand	Know we don't know or understand
Mistakenly think we know and understand	Unaware of what we do not know or understand

Knowledge (vertical axis label)

Awareness

Figure 1.2 Knowledge matrix

In the "lean" improvement world, we talk about push vs. pull. Do organiza-tions push their products on their customers, or do customers pull produc-tion so we make what the customer needs precisely when they need it? We can apply this same thought to leadership. Do your team members "pull" what you need to do as a leader for the performance and development of associates, or are you typically pushing them to get things done?

You can solve this conundrum and learn to see things currently invisible to your eyes. But it will require some practice and some patience.

If we were to look at this situation using a simple knowledge matrix, it might look like Figure 1.2.

There are things we know and understand as a leader; there are also things most of us are aware of that we do not know or understand. Typically, the latter is where we focus on our learning and development opportunities. The danger areas are in the bottom two quadrants:

■ We mistakenly think we know, or
■ We are unaware of what we do not know.

This book should help the reader become more aware of mistaken beliefs and share a few insights on becoming more aware of what they currently do not know. In Chapter 6, we focus on changing our habits.

Is There Really a Leadership Crisis?

There are many things written on "leadership." If you do an internet search, you will get over 830,000,000 hits, and on leadership books, more than 13 million. So much has been written, yet leadership problems still exist.

A survey report from the World Economic Forum[2] (WEF) shows that a startling 86% of respondents to their Survey on the Global Agenda state, "We have a leadership crisis in the world today." A finding that isn't at all surprising given the challenges faced in today's world. The WEF report further states, "It's not enough to just be inspirational; the best leaders know they must mediate, listen, and include the opinions of others before deciding. Execution, team building, and delegation are key, as is the ability to remain positive in the face of adversity."

Gallup, Wyatt, and other firms that measure "employee engagement" have shown in their research *the number one reason people cite for not being engaged at work.* It's not because the CEO isn't supporting improvement activities. The number one reason people say, "I am not engaged" is *because of the behaviors of their direct boss!*[3] So clearly, based on the information in the WEF report and from the employee survey results, we do have a leadership problem. *People leave managers, not companies.*[4] So, we have met the enemy of effective team performance and unfortunately, it's us (Figure 1.3)!

Leaders in the middle of an organization have a tremendous amount of leverage. Consider this simple fact: First- and second-line leaders directly touch 80% of the people in their organization.[5] They have a tremendous amount of influence and more power than they might realize.

I've never met a CEO who did not want their organization to get better. They all want improvement. However, given that many of them are blinded by their beliefs about how good the organization is, they often do not see the tremendous improvement potential that exists. If the actual reality is that most CEOs are not actively engaged in driving improvement, then perhaps it makes sense to focus on the individuals who do indeed directly touch most people within an organization.

Figure 1.3   We have met the enemy of effective team performance, and unfortunately, it's us!

Source: © Okefenokee Glee & Perloo, Inc. Used by permission. Contact permissions@ pogocomics.com

If you want the work by your team and by your peers to be more meaningful, then experiment with the four foundations described in this book and act. *Learn to see the invisible that likely exists in your environment.*

Leaders in organizations that are highly effective with their improvement practices lead in a way that differs greatly from most organizations. Leaders in this type of environment change the way they behave. It permits them to see things they likely would have missed in their former way of operating. It changes the way they look at their world. And that is always a pretty cool experience!

Gary Peterson is Senior Vice President at O.C. Tanner, a leader from whom I've learned over the years. He recently posted a YouTube example of seeing something new.[6] Gary was visiting the Acrylic Team and questioned why they had moved a station out of their production cell. He was pretty sure this was a waste, as it added travel time to get to the new location.

However, with his new behavioral norm, he asked a question and withheld his judgment. The team explained this station was more of a warehouse function for their group that took up space in the value-adding part of their work cell, where they made the product. Looking at this as a warehouse totally changed Gary's mindset, and he realized how the team benefited from this change. His willingness to lead with a degree of humility and ask a question allowed his eyes to see something he had previously missed.

Beliefs, Assumptions, and Habits Can Blind Our Eyes

Believing you are better than you are (as an individual, a team, or an organization) blinds you to seeing the actual reality. When leaders operate from that perspective, they unconsciously see things that reinforce those beliefs and cannot see things, even though they might be standing right in front of their noses, that conflict with those beliefs. It takes some effort to set your assumptions and beliefs aside and learn to see what is actually happening, to learn to see current practices that inhibit improvement. Here are three common beliefs/practices of many senior leaders:

1. **We Are World-class or We Are the Best at What We Do**: While many leaders like to believe their products, services, and organization are "World-Class," only a handful truly operate that way. It's easy to see why people want to believe this. It means you are pretty good at what you do. It suggests you are better than just about everybody else. And it suggests the way you currently operate as a leader is very good. Who wouldn't want to believe this?

    Unfortunately, this belief can be harmful. What if you are not "the best," but think you are? What if your beliefs and assumptions about how work gets done inhibits your ability to see what is really happening? And even if you are very good at what you do, how might the mindset of "we are great" impede getting even better at what you do? How might that belief inhibit doing a better job of leading people?

    Things lay hidden from our eyes in our day-to-day work environments with the way we lead and the way work gets done. One of my friends tells a story about how his organization did a workaround for over 20 years. They were building a large piece of equipment and routinely the operator had to grind a component part so it would fit. Their daily workaround became so common of an occurrence that both the people doing the work and their managers could not see the wasted

Figure 1.4   There is an elephant in the room?

effort that was taking place every day. Things we don't see literally surround us. And the further a leader gets from the people who do the actual value-adding work, make the physical product or deliver the actual service, the easier it is to be blinded by beliefs and assumptions that get made about the way work gets done (Figure 1.4).

People need to *learn how to look, to see the actual reality,* rather than finding things that reinforce their current beliefs.

2. **People Are Our Most Important Asset**: There is a lot of conversation, especially in the organizational improvement world, about *"respect for people."* What does respect mean? A typical definition states, "due regard for the feelings, wishes, rights, or traditions of others."[7] Are people typically respected in most organizations today? Unfortunately, the answer is they are not; that is not the way most organizations lead. We will share a few statistics on this shortly.

This is an old story and a great example of what can happen when people get treated with respect. The Toyota Motor Company and GM started a new joint venture in the early 1980s. They called it the New United Motor Manufacturing Company (NUMMI). Toyota agreed to take over the management of the worst quality performing plant that GM owned. GM wanted to learn how to profitably manufacture compact cars and Toyota wanted to learn if their management system would work in America. Under GM leadership, absenteeism would run 20% to 50%, and workers would sometimes sabotage cars by intentionally leaving loose bolts or soda cans inside car panels. Toyota was required to work with the same unionized workforce.

Within two years, the NUMMI plant went from GM's worst-performing plant to its highest-performing plant in quality output. How did they do that? Certainly, there was an improvement in some of the manufacturing processes, but mostly Toyota treated the employees with respect. They trusted employees to do the right thing, and when employees faced obstacles in getting work done, leadership immediately stepped in to help resolve the issues. When employees realized leadership really cared, attitudes and, most importantly, behaviors changed.

3. **Habits Are Comforting, Help Us Get Something Done Quickly and Require Little Thought**: Once you develop a habit, your brain defaults to automatic. You do the activity with little thought. Driving to work, you certainly look at your surroundings, but how often do you realize you have been going for the last 20 minutes and recollect little of what you saw? You are both somewhat observant and operating in a dreamy state without giving a lot of thought to what you are doing.

Sometimes, our habits impede our ability to see the actual reality of what is happening in our work environment. You haven't had a safety issue in over five years. When you do a walk, you expect everything to be OK, so your mind wanders, and you think about what you need to do when the walk is done, or you wrestle with the problem that came up just before your walk.

Bud Tribble coined the expression "reality distortion field" to describe Steve Jobs' ability to positively encourage his team to accomplish seemingly impossible goals. It is an expression that sounds to me like it should have come from a science fiction movie. But when senior leaders hold the first two beliefs, they create a "*negative* reality distortion field." Leaders hold beliefs that are out of touch with the actual reality.

It's a challenge for mid-level leaders working for out-of-touch senior leaders, but if you are going to stay in that position, it's your responsibility to help your team members grow. If you practice the four foundations described in this book, you will find positive ways to influence higher-level leaders in your organization. I'm not saying our world, or your organization, will ever be perfect. They won't! But you can make them both a better place.

Making Better Choices

A primary goal for a leader relative to "improvement" should be to create an environment where people make more effective choices more quickly. *We need to learn to see how some of our leadership behaviors may inhibit our personal and our team's growth.* But to transition to operate that way on a day-in, day-out basis is a challenge. You can't do this while sitting in your office or at your desk and *assume* you know what is happening. It requires a more activist approach where leaders go to where people are doing their work, find the facts, and build consensus to make better decisions and achieve goals. It requires deep reflection and learning to see the subtleties of how your organization, department, or team operates. Don't make dangerous assumptions about things you only know from a distance.

While we all develop habits for things we repeatedly do, it's essential to challenge them now and then, to ensure they are still appropriate in terms of how we should behave and how we should be present with our whole mind for what is happening at that moment.

Learning to See the Invisible and making better decisions requires changing your behaviors, and you will need time to practice becoming highly effective. Sports champions put in thousands of hours of practice. When we see them perform, it looks like they are natural athletes, which must come easily to them. But that is not the actual reality. *Professional athletes practice their skills repeatedly.* Their sport is what they do. They improve and make it look easy because they worked hard at getting better.

If you are a leader, shouldn't you be doing the same thing with your leadership skills? *Finding effective ways to practice getting better and better at leading is an important responsibility!* Learning to see what might be invisible to your eyes today can help you focus on where and what to practice.

Making better choices requires overcoming fear, both by leaders and by associates working within the organization. Leaders need to overcome multiple fears:

- Admit you don't know – intellectual humility by the leader creates a space for learning.
- A willingness to reach out and trust the people you work with to do the right thing – this is very frightening to many people (it was to me).
- Even when you know the answer, have the patience to allow your subordinates to gain experience – let them learn firsthand through their experimentation to achieve deeper learning.

Making better choices requires trust and vulnerability, where people are comfortable interacting with one another and respectfully asking challenging questions. It also requires people to improve their critical thinking skills, to let go of biases that inhibit their abilities to see reality as it really exists, and to see the positive power that comes from nurturing and growing capabilities. There is a great quote from *Ted Lasso*, the soccer coach in the AppleTV program by the same name, "Our choices show what/who we truly are, far more than our abilities."[8]

What Do People Want from Work?

I've had opportunities to travel worldwide and work with companies in many countries. One thing that is common to people everywhere is that *people want to feel they matter*. It doesn't matter if we are talking to home healthcare providers, an assembly line worker, a software developer, a machinist, or a CEO; they all seek:

- Respect from their boss, their co-workers, and peers with a sense of dignity.
- Meaning in their life and to live for something bigger than self.
- Pride in what they do and feel they make a meaningful difference.
- Opportunities to discover their capabilities and for self-development.

Fair compensation is desirable, but more is needed to inspire passion. Meaning can undoubtedly come from activities outside of work. The points listed above make a job more meaningful and differentiate how people feel about their direct boss and their organization.

Where Does That Leave Most Managers Working Inside Organizations?

Leadership starts with self. How capable are we, and how forgiving are we of our self-leadership? Do we have the humility to learn how to become better? If you cannot effectively practice self-leadership, where you consistently behave in a manner that is appropriate for the situation at hand, then it is quite difficult, and perhaps impossible, to lead others in a highly effective way.

Your direct boss or your CEO may not be a role model of excellence, but more importantly, what type of leader do *you choose* to be? If you wish to be on a championship team, what are the actions, routines, and leadership behaviors you, your teammates, and your support team need to practice to make it happen?

I can relate to the difficulty of doing this well. I've talked about "change" and performance improvement for over 30 years and like to see change happen. But I have a change problem. The things I want to change are in areas where it's easy for me to make a shift or perhaps deal with an immediate concern. But what if the change is hard to do, requiring discipline over a prolonged period? If I don't see an immediate need or if I don't want to work that hard to make a change, I'm as stubborn as they come. I doubt anyone reading this material is more stubborn than I am; if you don't believe me, ask my wife. She will attest to my pigheadedness.

But what if some of the things I don't particularly want to change are important to do? Then, I have a problem that will make it difficult for my team to operate at a champion level of performance. Perhaps readers of this book suffer from a similar problem. Relative to leadership capabilities, it is *unlikely* you are a completely dysfunctional leader. It is much more likely you are an average leader doing an OK job. Average is where most people operate. It is an acceptable, but not outstanding, level of performance.

If that is the case, then to achieve a "champion level of performance as a leader," we need to develop practice and feedback routines to help us get better. *Each chapter of this book should help people see things their eyes currently miss, develop the discipline to practice getting better daily, and use positive reinforcement and feedback to pursue higher levels of performance, to improve how we lead and how our team and our organization performs.*

Interestingly, we are not likely to make this leap by merely focusing on, and improving, our weaknesses. Working on our weaknesses is incremental improvement; weaknesses are things we already know we should do better.

But it requires something more if you want to become excellent as a leader. You may find it more important to focus on your strengths, scaling them up or dialing them back if you are doing too much of a good thing.

If you practice the above four foundations with humility, you will learn more about yourself and develop a mindset to improve your leadership capabilities and effectiveness. You will see things tomorrow that you do not see today, and you will touch others who will also learn and grow.

One key trait common to every highly effective organization we visited was the leaders' ability to change their perspective and look at their world through fresh eyes. A willingness to say, "I don't know the answer." Or perhaps, "I don't know the *only* answer." These leaders rarely tell someone how to do something. Instead, they ask good questions and help people learn how to do it.

Try It; You Might Find You Like It

I encourage you to take this journey. It's crucial to your team, your organization, the world at large, and the families of people who work with you. Repeating a thought shared at the beginning of this book, "Better organizational leaders can help the people around them to grow, to develop their capabilities, and to become the champion leaders of the future, thus helping their organization to grow in a meaningful and positive way" and indeed make the world a better place.

I certainly believe it can happen; I've seen it, I've done it, and so have others. You have an opportunity to become a better person and directly help the people around you to do the same. I hope you will accept the challenge.

Notes

1. AME, "Annual Conference Keynote talk." November 1, 2018.
2. https://www.weforum.org/agenda/2014/11/world-2015-faces-leadership-crisis/.
3. http://www.gallup.com/businessjournal/182321/employees-lot-managers.aspx.
4. https://www.gallup.com/workplace/232955/no-employee-benefit-no-one-talking.aspx.
5. https://hbr.org/2011/05/the-frontline-advantage.
6. https://www.youtube.com/shorts/gGf0HLvaoYo.
7. Oxford Languages.
8. Coach *Ted Lasso* in Season 2, Episode 12 – Apple TV.

Chapter 2

Reflection

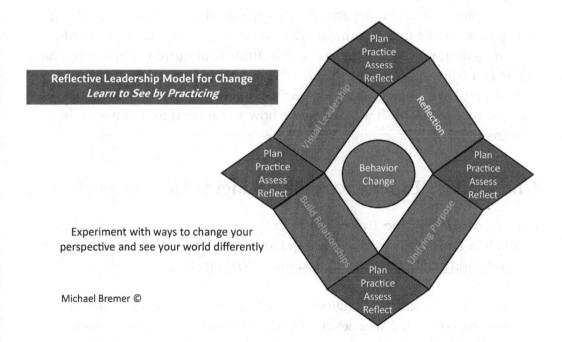

Reflective Leadership Model for Change
Learn to See by Practicing

Plan Practice Assess Reflect

Plan Practice Assess Reflect

Behavior Change

Plan Practice Assess Reflect

Plan Practice Assess Reflect

Visual Leadership

Reflection

Build Relationships

Unifying Purpose

Experiment with ways to change your perspective and see your world differently

Michael Bremer ©

Curiosity

It takes a curious mind to learn to see tomorrow what your eyes/ears/senses miss today. Do you have an interest in learning? Do you seek to develop a deeper understanding of the people around you to help them, and you, to grow? Do you have an interest in improving your leadership skills? If yes, it starts with changing your perspective and, as our title suggests, "learn to see things tomorrow that perhaps you do not clearly see today." We will talk a

DOI: 10.4324/9781003495284-2

lot in this chapter about the importance of reflection, slowing down to consider what is happening in your life and your day-to-day experiences. This is not an original thought. In Romans 12:2 of the Christian Bible, there is a line, "do not be conformed to this world, but be transformed by the renewing of your mind, that you may prove what is good." Reflection is a great way for us to begin the learning process.

There are multiple ways in which you can practice reflection to change your perspective; we share a few thoughts/tools/techniques at the end of this chapter in the Actions to Practice section.

Most leaders, in my experience, do not need a radical makeover. We are looking for subtle adjustments, which are sometimes the most difficult to see. Because, in all likelihood, much of what you are doing is OK. It's just not the best you can be.

To inspire your team, department, or organization to make meaningful changes, you must first determine what needs improvement. As it stands, you are getting precisely what you put in from your current ways of operating. If you want a significant shift for the betterment of your team, then you should start by changing your perspective. Reflecting on your status quo is the best way to establish a baseline for how to proceed in making future adjustments.

One Leader's Experience with Changing Her Perspective

Jess Orr worked for the Toyota Motor Corporation for several years and is currently a Performance Excellence Practitioner at WestRock, a large paper and packaging company. She shared the following story:

> As a newly graduated engineer, they trained me to be an expert in solving problems. Whenever I faced a challenging task in my job, I would gather the facts, put my head down, and come up with a solution. I was moderately successful in the early stages of my career. I prided myself on being known as an analytical and creative problem-solver with a reputation for achieving results.
>
> However, the way I did my work limited how effectively people sustained it after I left the project. For example, an innovative piece of manufacturing equipment, I had helped develop and implement stood idle on the production's corner floor. I became frustrated at the progressive decline in the results of my work after my initial

involvement and wondered if the fact that I had worked on these projects primarily by myself was part of the problem.

After a few years, I took a quality engineering job at Toyota in Georgetown, KY, which produced the Camry, Avalon, and Venza vehicles. It was here that I was first introduced to Toyota's A3 problem-solving process. They taught me that forming a team was one of the prerequisites to working through an A3. These people would function as subject-matter experts because of their technical knowledge or direct experience with the problem. Although I adapted my approach to include others in the process, I still did most of the work. Involving others was more of a method for ensuring their buy-in to sustain the solutions, and not as a direct means of solving the problem itself.

After leaving Toyota, I joined a large paper and packaging company as a full-time continuous improvement practitioner. During my first few projects, I made much more effort to engage process owners and stakeholders in finding ways to improve the work. I recall one time when I had my team brainstorm solutions to reduce costs in their area, and then had them vote on ones they wanted to pursue further. I was certain their Top Idea was not likely to be successful. However, I held my tongue and allowed the team to pursue the improvement. The improvement worked so well that it saved over $200,000 per year. Results like these and others convinced me that engaging others was more than a courtesy and a way to ensure their buy-in; instead, it was a way to find solutions better than any I could come up with by myself. *I gradually found my role shifting from less of a leader driving change to more of a facilitator of others to improve their own work.*

A recent experience both reinforced and expanded my view of the power of this approach. They had asked me to lead one last effort to save one of our facilities that had become unprofitable. I remember how, shortly after being assigned the project, I placed my head in my hands, at my desk, as I felt the total weight of the jobs of the facility's 200 employees on my shoulders. I knew that just one person could not overcome this obstacle – but if we could leverage the minds and efforts of those 200 people, we might stand a chance.

I threw myself entirely into facilitating employee-driven continuous improvement. Knowing the leadership at the plant would be critical to success, I focused first on developing them on the principles of servant leadership. During our many production floor walks, we sought opportunities to help employees solve immediate problems and asked for ideas to improve their work. We enabled the associates to implement the ideas themselves, providing guidance and support as needed.

Slowly, the conditions improved. The results were enough to convince our leadership to keep the plant open. When they visited to see the improved condition, we had the production team members themselves explain the ideas they had implemented. They took complete ownership of the improvements because they had driven them, and I knew without a doubt that the results gained would be sustained long after I left. As I watched the team members with pride, I realized my fulfillment would come not by trying to leave my mark on the world but through enabling and empowering others to make theirs.[1]

Reflection – Things to Think About

Reflect on Your Current Situation

"We already do that!" is one reason why it is so difficult to learn. Because the situation is *familiar,* we think we *understand* it and fail to look for deeper, more profound meaning. One of my favorite quotes is from Satchel Paige, a phenomenal pitcher in the Negro Baseball League for many years. He once said, "It ain't what you know that gets you in trouble. It's what you think you know that just ain't so."[2] Dig deeper to look at *"what you think you know that might not be so."*

Jess's story is a good example of stepping back and looking at what is happening. There is reality as we wish it existed. It usually revolves around our intentions. In Jess's story, she wanted the project work she did to be outstanding, and she wanted it to be used for a long time. The actual reality is a little different. Given the "good intentions and assumptions" we make, the "actual reality" is hard to see. When things don't work out as we expected, sometimes there is a tendency to blame others or do a workaround. It takes effort to learn to see the invisible. Rather than going faster, we typically need to slow down and reflect. We need to see the world around us in a different light or a different angle (Figure 2.1).

Jess's story is also good because it shows how long it takes for us to learn new behaviors. She was moving in the right direction as she progressed in her approach, but it took several attempts to not only learn but also to "believe" and trust new ways of doing work. Mentally, we expect this shift to be like a light switch. Yesterday we were this way, and today we are this brand-new shiny thing. Unfortunately, that is not how life works.

It might seem that Jess was resistant to change, but she was experiencing the normal growing pains of learning and applying new behaviors. Her initial approach was based upon years of experience, and naturally, it was difficult, as it is for all of us, to embrace change. Ultimately, she was successful! You shouldn't just simply abandon what works, but you shouldn't also hold on to it so tightly that you cannot learn something new and improve your approach.

Figure 2.1   Changing perspective

The beauty of what Jess did was she progressively learned from each attempt at engaging people and adapted her approach accordingly. In every subsequent experiment, she trusted people more and learned how to better lead them along the way to become a more effective leader.

Change starts with you, not with others. What is your current state? What do you need to change to lead more effectively? The challenge is to avoid only learning this via the "school of hard knocks – experience" and instead progress more quickly. Reflection can help with acceleration.

Change Your Perspective *Before Changing Others*

In his book *Principles: Life and Work*[3] and related podcast, Ray Dalio, founder and CEO of Bridgewater Associates, said, "I needed to let go of the *Joy of Being Right* and replace it with the *Joy of Learning What Is True.*" He realized that the way he was leading was an inhibitor to future growth. He then hired the most thoughtful people who disagreed with his point of view and sought to learn from them. He wanted to become a better listener *"to see things through their eyes … and to reflect on their reactions to his thoughts.* Going from seeing things in black and white to instead, see the color to see the subtleties of what was really happening."

Many of us have a pretty good idea of what other people need to do to improve. We all have great insights to see the shortcomings in the way other people operate. And probably a few of us have said, "If only Mary or Joe or that department would *(fill in the blank)*, my life could be so much better, my job could be so much more effective."

Highly effective leaders use a different approach. They consistently say, *"In order to elevate the way I lead and take it to a higher level of effectiveness, I need to change the way I think and operate."* 100% of the highly effective leaders we studied said something very similar. They started with "What do I need to do to lead more effectively? What do I need to do? What does our department (team) need to do? What does our company need to do?" that I might influence, before focusing on other people, other departments, suppliers, or customers. There is power in gaining new insight, and it will often result in a new sense of freedom!

This certainly suggests our perspective needs to shift if we plan to make a major change. It starts with a desire to see our world differently. A relatively famous quote is attributed to several people: "the definition of insanity is to continue doing the same thing over and over again and expecting a different outcome." Our habits and ways of operating need to change if we expect something different to happen.

You might wonder how much "shift" is needed? And the simple answer is probably more than you think, in terms of *perspective shift (i.e. your ability to see your world differently, to see it through new eyes).* When Taiichi Ohno's disciples thought they fully understood the Toyota Production/ Thinking System, he would ask them another question, and they would realize that they still had much to learn. So it is for us as leaders. Learning how to ask meaningful questions and maintaining a degree of humility sets the stage for all of us to become more effective leaders.

Defining Why You *Need to Change Is Important*

What if a well-informed, trusted authority figure said you had to make difficult and enduring changes in the way you think and act? If you didn't, your time would end soon – a lot sooner than it had to. Could you change when change really mattered? Most of us would probably say, "Of course we could!" But the reality is, most of the time, people, even when faced with death, do not change, according to a Fast Company article by Alan Deutschman, "Change or Die" published in 2005.[4] He later published a book with the same title (2007).

Many patients with bypass surgery could avoid the return of pain and the need to repeat the surgery – not to mention arrest the course of their disease before it kills them – by switching to healthier lifestyles. Yet very few do, as high as 90% in some studies. Why? They also address the answer to this dilemma in the article.

Dr. Dean Ornish, a professor of medicine at the University of California at San Francisco and founder of the Preventative Medicine Research Institute, realized the importance of going beyond the facts. He says, *"Providing health information is important but not always sufficient, we also need to bring in the psychological, emotional, and spiritual dimensions that are so often ignored"* [emphasis added].

Researchers took 333 patients with severely clogged arteries. They helped them quit smoking and go on Ornish's diet. The patients attended twice-weekly group support sessions led by a psychologist and received instruction in meditation, relaxation, yoga, and aerobic exercise. The program lasted for only a year. After three years, the study found, 77% of the patients had stuck with their lifestyle changes and safely avoided a new bypass or angioplasty surgery.

Why did the Ornish program succeed for people in the study while the conventional approach failed? For starters, Ornish *recasts the reasons for change. They shifted peoples' perspective.* Ornish said:

Traditionally doctors had been trying to motivate patients mainly with the fear of death, and that simply wasn't working. For a few weeks after a heart attack, patients were scared enough to do whatever their doctors said. But over a short period of time, their fear would diminish, and they'd go back to their old habits.

The patients lived the way they did as a day-to-day strategy to cope with their emotional troubles. Ornish says, "Telling people who are lonely and depressed that they're going to live longer if they quit smoking or change their diet and lifestyle is not that motivating, *Who wants to live longer when you're in chronic emotional pain?"* [emphasis added].

So instead of trying to motivate them with the "fear of dying," Ornish *reframes the issue.* He changes their perspective to inspire a new vision on the "joy of living" – convincing them they can feel better, not just live longer. That means enjoying the things that make daily life pleasurable, like loving someone else or taking long walks without the pain caused by their disease. *"Joy is a more powerful motivator than fear,"* he says. *It very much shifts the person's perspective on life* [emphasis added].

"Changing the behavior of people isn't just the biggest challenge in health care. It's the most important challenge for businesses trying to compete in a turbulent world," says John Kotter, a Harvard Business School professor who has studied dozens of organizations during upheaval: *"The central issue is never strategy, structure, culture, or systems. The core of the matter is always about changing the behavior of people"* [5] [emphasis added]. In today's world, consultants especially like to use the word *culture.* The core of the matter is much more simplistic; *it deals with the way we and others behave.* Do we behave in the way we are most comfortable operating, or do we seek to identify the behaviors most appropriate for the challenges facing our team, our changing workforce, our department, our company in tomorrow's environment?

As individuals, we may want to change our own styles at work – how we mentor subordinates, for example, or how we react to criticism. Yet often, we can't; we get trapped by behaving in ways that are comfortable, behaviors that in our past used to work. Reflection can help us see that perhaps those habits are no longer appropriate for the new situation.

Kotter, Deutschman, and Ornish have all hit on a crucial insight to break out of the trap. *Behavior change happens mostly by speaking to people's feelings.*

Creating a meaningful definition of "why you" need to change will probably require working on your mindset, working on how you look at the world, how you look at your work, and how you look at your role and responsibilities for leading others. We come back to this concept of "why" in the chapter on Unifying Purpose (Chapter 3).

Establish the Mindset

Most of us operate with the mindset that things are going well, everything is (largely) OK. If you are truly interested in continuous improvement, as was Taiichi Ohno at the Toyota Motor Company, you need to change your mindset to *everything is a mess, and we could do much better!* Establish a mindset where everything you do is an experiment, and you are constantly learning about the results and seeking to find *a better way.*

We behave in certain ways as a leader because we believe it is the right thing to do. But what if you could change your perspective and see your world, your role, differently? Might you then desire to change the way you lead?

Be the change you wish to see, instead of trying to get others to change. Most leaders:

1. Set direction with challenging targets.
2. Learn to more effectively see problems, abnormalities, waste, and opportunities.

Outstanding leaders purposefully go further and increase trust levels with employees and between cross-functional work groups. They seek to:

1. Create an environment where team members feel safe to point out problems/issues.
2. Teach/coach associates to develop their ability to perform, fix and improve their processes and their critical thinking skills to accomplish meaningful performance targets.
3. Have the tenacity to stay the course, and balance that drive with a humility that permits them to stay in touch with reality as it actually exists.

Story of a Mindset Shift Change

A good friend shared this story. I've known this individual for over twenty years, and always thought he was a bright person and a good leader. But he was always a man in a hurry to get things done. The transformation of his leadership style was a bit of a challenge, but not at all surprising in the way he handled it.

> I was naïve early in my career. I've been a loyal employee to my employers and tried to play the game as seemed appropriate for the organization. I did not jump ship from company to company. Started my career at a large Canadian company (14 years) which was acquired by a global U.S. Company (15 years). And I'm now at my new company (a Fortune 100 for 13+ years).
>
> I always felt when I was in Rome, I should do what the Romans do and find a way to be effective within the respective company's culture. At the U.S. Company, it was a full contact sport. If you wanted to survive, it was the opposite of being a servant leader and I was relatively successful at doing it.
>
> When I followed my boss to my new company, I was quite excited. My boss and the new CEO wanted me to lead our lean improvement efforts globally. "I was quite good at leaning things out." Within six weeks of my starting, I was called into the VP of HR's office. I thought things were going great! He informed me that things were not working out! There was one Division President in particular who wanted me to be fired. This caused me to reflect on my behaviors; I did a lot of soul searching.
>
> Previously, I felt I was a maniac on getting results. I was aware of lean principles and of the importance of people, but mostly I was getting results by driving them, I was not acting as a leader coach.
>
> So, pretty much in a "cold turkey fashion," I changed that day. It made a big difference in my career. A couple of months later I was doing a Gemba Walk through an operation with the Division President who wanted to fire me. It was extremely uncomfortable. At the end of the walk I asked, "Could we talk?" We went to a meeting room and closed the door and I said, I realize now I was like a bulldog at the start … I was wrong, and I want to apologize. There were many good things going on that I was not taking time to recognize, nor was I listening.

Figure 2.2   Which version of yourself will show up for leadership today?

At the end of the conversation, the Div. VP shook my hand and thanked me and accepted my apology. Two years later I went to work for him on an assignment that was very successful. I learned you can recover if you take responsibility for your actions, be accountable and show a willingness to change.

Hold yourself accountable for bringing out your "best" (Figure 2.2).

It has been interesting to watch his evolution to lead with more humility and a much greater willingness to listen to alternative perspectives. What most interested me in this story was his willingness to stop and listen, to reflect on what was being said, and then, most impressively, his willingness to take responsibility for his actions and to change his behavior. *This is never easy to do, but the rewards are considerable, especially when you realize how many more people you can positively touch with the work you are doing by improving the way you lead.*

I don't want to focus much in this book on the character qualities of effective leaders. Much has been published on that subject. However, Google did a very interesting study that is worth a quick review.

Google's Traits of Highly Effective Leaders

Google launched a comprehensive research project named Project Oxygen in 2008 to pinpoint and quantify essential managerial behaviors that significantly influence employee motivation, group engagement, and team performance. This initiative focused on behaviors typically exhibited by leaders managing small- to medium-sized groups.

The project originally identified eight key behaviors for effective leadership; they later added two additional items to the list.

1. **Be a Good Coach**: Conversations are important! Provide detailed, constructive criticism, while ensuring a balance between positive and negative feedback. Conduct regular one-on-one meetings, providing problem-solving approaches that leverage employees' unique strengths. Coach more by asking questions rather than providing solutions.
2. **Empower Your Team – Don't Micromanage**: Balance between allowing autonomy and being available for guidance. Assign challenging tasks that enable the team to address meaningful issues.
3. **Create an Inclusive Team Environment Supporting Team Members' Success and Well-Being**: Get to know your employees personally, and acknowledge their lives beyond the workplace. Welcome new team members warmly and facilitate their integration into the group.
4. **Be Productive and Results-Oriented**: Emphasize the team's objectives and how each member can contribute. Prioritize tasks and eliminate obstacles hindering progress.
5. **Be a Good Communicator and Listen to Your Team**: Practice bidirectional communication – listen actively and share information. Conduct comprehensive meetings that convey clear messages and objectives. Help the team understand their interconnectedness. Promote open discussions and address your employees' issues and concerns.
6. **Support Employee Career Advancement**: Understand each employee's career aspirations. Discuss potential career paths and necessary steps for advancement. Identify and address skill gaps to facilitate progress.

7. **Maintain a Clear Vision and Strategy for the Team**: Keep the team focused on goals and strategy, even in challenging times. Engage the team in developing the team's vision and working toward it.

8. **Possess Essential Technical Skills to Advise the Team**: Be hands-on and work alongside the team when necessary. Comprehend the specific challenges of the work. Ensure you are approachable, consistent, inclusive, clear in your expectations, and fair. Effective managers don't just communicate; they connect with their team.

9. **Collaborate Across the Organization (Google)**: Actively seek to cooperate across the organization, developing relationships with different groups. Don't operate in a silo, just focused on your metrics. Realize customers get served by interdependent processes that run across organizational silos. Maximize overall process performance, not just your silo.

10. **Be a Strong Decision-Maker**: Be decisive in your decision-making. Don't waffle or procrastinate. Communicate the reason "why" behind your decisions.

The above traits are listed in priority order for Google's environment. The sequence or even some items on the list may differ for a company outside of the technology industry. But it is a reasonable list of traits and analytical research supports it. *This is a good list for reflection, especially since the key behaviors from their research primarily describe leaders of small- and medium-sized groups and teams; this is especially relevant to first- and second-level managers.*

Change Your Perspective

Everyone I've spoken with who had the opportunity to work with a true Toyota Sensei will say one thing in common. The experience changed the way they think and the way they behaved. Toyota Sensei, especially those who worked under the guidance of Taiichi Ohno, helped these leaders to develop a deep understanding of highly effective organizational/leadership improvement practices. They changed the way they thought about improvement. They changed the way they behaved as leaders. Toyota is still a "role model" for organizations with highly effective improvement practices. Mike Rother, who studied Toyota as much as anyone, recently indicated that he didn't get it, but now has a better handle on it. *I believe that is our goal. We all need to strive to get a better handle on it and behave more appropriately.*[6]

There are many ways to change the way we see the world and change the way we think. There are free online personality assessments like Meyers Briggs or DISC that help you understand how you think. Some companies use 360° feedback, which can provide insights. Using the typical lean analytical tools like process mapping, value stream maps, simple time analysis, etc., helps us to gain insights. Reading a book can change how you see the world, how you see your job, or how you see yourself. Hopefully, your education also changed your perspective as you learned things you did not know, and things you did not understand.

There are three areas where you should focus on gaining a new perspective, and they are listed in a priority sequence. Start with #1:

1. **Leader Reflection:** Start with yourself as is emphasized throughout this book.
2. **Team Reflection:** Help your team see their world through fresh eyes.
3. **Process Reflection:** Find ways to make it easier for your teammates, your peers, and your internal customers/suppliers to see what is happening with cross-functional process performance.

Ideally, the *Actions to Practice Executing in Each of the Four Foundations* will give you a few ideas on practical steps you can take to become a better leader. Start with reflecting on your current situation and then practice changing your perspective. *You can learn to see your world with fresh eyes, and learn to see what is currently invisible!*

The actions listed below are not in sequence. They are options to consider, and your choices will vary based on your current situation, your current leadership style, and your personal, team, and company objectives. Many of the actions described below can influence all three areas of reflection (self, team, process), depending on the perspective taken and the situation.

Scan this list. What jumps out as a meaningful, potential action? What action can help you take one step and then another? People climb mountains one step at a time. While we all might wish we could jump right to the finish line … it is instead a learning journey. Experiment with these actions, learn how to build upon this foundation. *These tools can be used by an individual leader, a team, by a department, a leadership team, or could even be used across a company.*

Actions to Practice – Reflection

Define the Desired Behaviors

This action is so simple, I can't believe that everyone doesn't use it to practice reflection. These first two tools can also be used for improving team effectiveness, and *if you are brave enough, you could also try using them regarding your role as a leader.* What do your team members think you should keep doing, stop doing, and start doing? This would be a simple way to quickly get a 360° type feedback from your team members.

   When people take part in project teams, the team will often define/agree to a few behaviors they want to follow. For example:

1. Start/stop meetings on time.
2. Only one person talks at a time.
3. Don't solve problems in the meeting.

Those are three behavioral norms we've seen many teams adopt over the past 30 years. They deal with normal problems faced by many teams. But the key isn't just agreeing to the desirable behaviors; the magic ingredient is finding a simple way to keep the individual team members accountable for practicing these behaviors. And it gets done through *two simple feedback mechanisms.*

Keep/Stop/Start

This is a little more open-ended. It could focus on a variety of things the leader/team/department/workgroup is doing.

- **Keep**: What should I keep doing that is driving us toward effective and efficient fulfillment of our purpose?
- **Stop**: What should I stop doing that is inhibiting our ability to effectively and efficiently fulfill our purpose?
- **Start**: What should I start doing to increase our effectiveness and efficient fulfillment of our purpose?

Keep/Stop/Start could focus on your role as a team leader or for target-specific behaviors for your team.

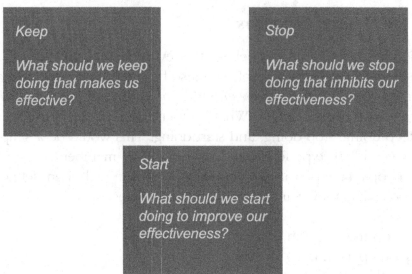

Figure 2.3   Keep/stop/start

We have always liked to do this using square Post-its and having people put the sticky notes on a flipchart or Whiteboard. Someone can then organize the feedback and do a readout at the end of a meeting or the start of the next meeting (Figure 2.3).

Measure Desired Behavior's Level of Effectiveness

This is more quantitative relative to modifying behaviors. Two or three behaviors should be identified that the leader/team/department/workgroup/ company feels would be more appropriate for the challenges. Remember, this is an experiment, so the best behaviors may or may not get identified on the first pass. Let's assume, for example, a team has identified these three behaviors as desirable for their leader to practice:

1. Asking good questions.
2. Listening without interrupting.
3. Looking at the process before fixing a problem.

You can quantitatively measure how well you perform each desirable behavior. Again, we like to do this using a flip-chart sheet, but it could also be

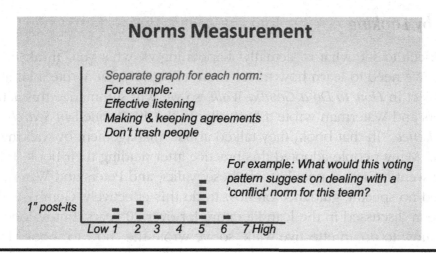

Figure 2.4 Norms measurement

done electronically. Create a bar chart for each desirable behavior. We use a 1–7 scale where "1" is awful and "7" is fantastic! On the flipchart, we list the ratings 1–7 along the bottom axis, and then people create a "bar chart" with one-inch Post-its™. You get something like the graphic above (Figure 2.4).

It would be best if you then debriefed the feedback. We always first ask, "Why did the minority vote give the lowest score?" In the above example, I would first question, "Why did the people who scored this behavior a 1 or a 2 score it that way?" Occasionally, you will get a real "gem" where one person has a unique perspective on the behavior being discussed, creating an insight for the rest of the group. This does not happen often, but when it does, it is precious. If you start with discussing the majority's view, you will never benefit from the minority's insight, as people will be more reluctant to share their thoughts. In the above example, the scores were bi-modal, with some people thinking things were going well and the other group not quite so satisfied. Seeing something like this posted on the wall or on a computer screen provides an opportunity for conversation and learning, so long as it is safe for people to discuss their viewpoints.

Once the leader or the department is practicing the new behavior effectively and sustainably, they can drop that behavior as a target and move on to the next one they want to do.

This is such a simple way to drive/guide behavior change. I encourage you to try it. *I've rarely seen this being done outside project teams, which is unfortunate.*

Learn by Looking

It is difficult to see what is "actually" happening vs. what you "think" is happening. We need to learn how to see more effectively. We wrote a lot about this subject in *How to Do a Gemba Walk,* so we will summarize this action.

Peters and Waterman wrote a bestseller in the 1980s titled *In Search of Excellence.*[7] In that book, they talked about "management-by-walking-around." Many people adopted this practice after reading their book. But as time went by, fewer people took these walks, and Peters and Waterman provided no specific guidance on how to do this effectively. Gemba Walks have been discussed in the lean literature for over 20 years. But no one told people how to do an effective walk, so we wrote the "how to" book. The typical literature said, "Go See, Ask Why, and Show Respect." I agree those are good things to do, but how do I effectively make it happen?

There are three key components to doing an effective walk:

1. **Define a Purpose:** Not surprised (I hope). Clearly, this is the first step prior to doing a walk. The purpose of a manufacturing first-line supervisor, a plant manager, a department manager in an office or R&D environment, or an outside executive will all differ. A few examples of this are included in the next chapter.
2. **Do the Walk:** This includes the common items of "Go See, Ask What, then Why, and Show Respect." I include quite a few examples in this book. Walking a stable process differs from walking an unstable process. An office walk in a customer service center will differ from a walk around a manufacturing operation.
3. **Debrief the Walk:** And also periodically debrief the walk process. *Three questions we should ask at the end of every walk:*
   a. Did we make any decisions?
   b. If yes, how are we going to communicate that decision?
   c. How are we going to follow up on progress?

You should also use those same three questions at the end of any meeting. If a decision is made, make sure people are agreeing to the same decision and what is going to happen next. That step, all by itself, can go a long way toward having better meetings. More information on this technique is in the Purpose chapter (Chapter 3) *under the Actions to Practice – Define a Unifying Purpose section.*

Yogi Berra once said, "You can learn a lot by just looking." But to learn by looking, one must *spend time learning how to look.* Offhand, that may sound stupid; anyone with normal eyes can see, but learning what to look for is not always straightforward. Think about doctors learning to diagnose patients or executives who are two levels away from the action. Our visual attention is easily tricked. The well-known video of a gorilla crossing between a team of people passing a basketball illustrates distraction from "the obvious." Viewers asked to concentrate on counting exchanges of the ball rarely spot the gorilla walking across the lot between the ball passes.

For example: Leaders often do damage when they try to fix problems several levels below their areas of responsibility. One reason contributing to this issue is that problems typically get filtered as they are shared going up the hierarchy. Lower-level managers feel it is their responsibility to fix the situation, so they are reluctant to burden their boss with all the details. The end result: Leaders several levels away from the problem often do not know what the real problem is. Contextual learning is very important, and understanding what is happening by seeing it creates new learning opportunities. That's why Toyota maintains the maxim of going to the physical scene of a problem before deciding what to do and teaching leaders how to see what is actually happening.

SIPOC

It is a very simple tool, perhaps so simple that it gets underutilized. This is typically used as part of process analysis, but *you can also create a SIPOC for your role as a leader.* Determine what data to collect by identifying the key stakeholders linked to your leadership activities, and identify what outputs are expected from your role as a leader (Figure 2.5).

1. **Suppliers**: People who supply information for the work you do.
2. **Inputs**: Specific inputs into your daily leadership practices (e.g. schedule, leader standard work routines, daily meetings, etc.).
3. **Processes**: What are the routine process activities you do as a leader (e.g. Gemba Walks, huddle meetings, etc.)?
4. **Outputs**: What results do you expect from your leadership activities (e.g. goals met, people developed, improvements implemented, etc.)?
5. **Customers**: Who are the direct personnel whom you are leading, serving, and/or influencing (e.g. team members, boss, peers, etc.)?

Review the map with all key stakeholders to verify.

Process Description:			Date Completed:	
Completed by:				

SIPOC Diagram

Suppliers	Inputs	Processes	Outputs	Customers
Who supplies the process inputs?	**What inputs are required?**	**What are the major steps in the process?**	**What are the process outputs?**	**Who receives the outputs?**

Note: Information is not linked across rows

Figure 2.5   SIPOC diagram

A *SIPOC* analysis should help you answer the following questions:

- What is your role as a leader? What purpose are you trying to fulfill?
- What are the inputs required to be successful?
- How can you tell if you are doing this effectively? What metrics show if you are on target, and what metrics drive improvement?
- What is the result you expect to see (may be different in the short, mid, and long term)?
- Most importantly, what do you need to improve (as a leader), and how will you know if you are effectively making it happen?

This tool can also be used for the work by a team, department, or business; it provides a great overview of the work being done and provides an opportunity to discuss "why" this work is important and how it fits within the overall chain.

Questions to Ask at the End of Every Year

The end of the year is a time that many people use to reflect on what happened during the last 12 months. Take time to consider your relationships at work, with your family, and with your community. Recognize that time is precious, and where you spend your time is a reflection to others relative to what you believe to be important. Avoid getting trapped by the day-to-day activities you do and make sure you give some thought to what living a meaningful life means to you. Most of the successful leaders I know seem to delight in uplifting the people around them; they grow by helping others to grow. This is important to do at work, at home, and in your community. Those groups only get better when people take the time to help make them better; it does not happen automatically.

- What did you accomplish that was hard to do? Did you take time to appreciate the accomplishment and thank anyone who helped you along the way?
- Who impacted your life during the last 12 months? How so? Did you thank them for the positive contribution they made to your life at work, at home, or in your community?
- How much time did you spend with friends or family?
- What did you accomplish in terms of self-improvement during the last 12 months at work, with family and/or for yourself?
- Thinking about family – what is important to your spouse, child, parents' life right now, and why is that important to them?

Looking forward to the coming 12 months:

- What might you do differently to have a more positive impact on the important people in your life?
- What would you like to do differently regarding the way you spend your time?
- Who have you not been spending time with who deserves more of your attention?

After reflecting:

- What is your decision on the one or two most important changes you would like to accomplish?
- How will you hold yourself accountable to stay the course for progressing toward your targets daily?

John Colby shared a neat trick at an AME conference. Don't ask, "Are there any questions?" Instead say, "Before we move on, I'd like to hear three questions."[8] You will get a different result.

These next several tools and techniques might help you to make better decisions as a leader or to guide your team in improving *process perfor-mance*. I realize this is a stretch for mid-level managers, but over the years when people have publicly posted visual metrics dealing with organizational processes, it has almost always positively influenced collaboration and alignment. We will return to this thought in the Visual Leadership chapter (Chapter 5).

Reflection Actions for Improving Processes and/or Decision-making

Looking at a process is different from reflecting on your personal leadership style or leadership actions. A middle manager rarely owns any of the processes with which they daily interact. Making improvements to a piece of a process without understanding the broader scope is often a colossal waste of time. Exceptions certainly exist, but the norm is when we only focus on the problem and not the process, the improvements are less likely to be sustained. What are some tools you might use (beyond process mapping) to improve processes and organizational decision-making?

Reframe the Problem

This entire book is basically about reframing – looking at the way you currently lead and then seeking ways to become a better leader. Carol Dweck, a Professor of Psychology at Stanford University, talks about a "growth mindset" vs. a "fixed mindset" focused on why people do or do not succeed, and what is within your control to foster a higher level of success. Simply speaking, what mindset guides your day-to-day behavior? If you believe your leadership qualities are already very good – a fixed mindset – you then seek to prove you are correct rather than learning from your mistakes. A growth mindset creates a powerful passion for learning. "Why waste time proving over and over how great you are," Dweck writes, "when you could get better?"[9]

In the earlier reference to Ray Dalio, he reframed his approach to learning. He moved from wanting to show others he was right to instead learning what is true by gaining a deeper understanding of other people's view of the truth. He reframed the way he thought and the way he learned by developing better listening skills.

Having a growth mindset helps when you seek to reframe. You become less worried about losing something and instead focus on something more positive, in this instance, personal growth.

In the May 2005 issue of the *Harvard Business Review*, Richard Tanner Pascale and Jerry Sternin wrote an article titled, "Your Company's Secret Change Agents." They share a story of an elementary school in the rural province of Misiones, Argentina, where student dropout rates were extremely high. The classic definition of the problem (commonly given around the world, in poor districts) was *"teachers were not paid enough, parents did not care, the facilities are lousy, etc"*[10] [emphasis added]. The teachers at this school reframed the problem of why students were dropping out. They negotiated "learning contracts" with rural parents before the beginning of each school year. In effect, the teachers were enrolling illiterate parents as partners in their children's education. As the children learned to read, add, and subtract, they could help their parents take advantage of government subsidies and compute the amount earned from crops or owed at the village store to make certain their parents were being treated fairly in these relationships. With parents as partners, students showed up at school and did their assignments. One year later, dropout rates in Misiones decreased by half.

Reframing was also used by Dr. Dean Ornish in "Change or Die." He switched the emphasis from don't do these things that will kill you to instead focus on the joys of living. A positive growth-minded perspective can be a powerful motivator.

One way to reframe is to see if a person, a team, or a department in your organization is doing a better job than the rest – what's going on? What are they doing to be successful? What can you learn from looking at where things are being done quite well? Is there a learning opportunity? This is a powerful change tool, and the learning experience can be both powerful and emotional. Obviously, there is also power in removing the filters (biases) we all carry and learning new "truths" which fits with Dr. Deming's idea of Profound Knowledge.

80/20 Rule (Pareto's Law) – What Should You NOT Be Doing?

Consider using the 80/20 Rule to figure out what is important. For example: If 20% of customers, products, or some other subset equals 80% of profits, then 80% of customers, products, or some other subset yields 20% of profits, and many of these are money-losing endeavors. Use this type of thinking to figure out what you should *not be doing*. Your most profitable customers are typically 16 times more profitable than customers in the bottom group.

How might you apply this thinking to your role as a leader? Eighty percent of the problems most likely happen with 20% of the work activities. Where does 80% of your time go on a daily basis? Is that where your time should go? Scrutinize the items on your "To Do" list, chances are very few of the items link to important issues. While we may take satisfaction in crossing a large number of (the smaller) issues off our task lists, the 80/20 rule suggests we should focus on the few, larger items that will generate the most significant results. In Stephen Covey's book, *7 Habits of Highly Effective People*, he called the key activities important but not urgent.[11] Make sure you know what is important over a longer period. What should you *Not Be Doing* on a daily basis that currently consumes your time?

Rule of 3 Alternatives

One of the first significant projects I undertook as a young man was to lead a project team evaluating a company-owned corporate research center. Beatrice Foods (a Fortune 30 Company at the time) had a research center that developed new products for a diversified set of independent businesses that Beatrice owned. Beatrice had over 440 separate businesses spread around the globe, including companies like Samsonite, Tropicana Orange Juice, Culligan, Meadow Gold, Dannon, etc. Most of the products developed by the research center were not picked up by any of the 440 profit centers. The R&D Center leadership would typically come up with an idea and then look for someone to adopt it (meaning provide more funding to further develop the concept).

Some of the senior managers wanted to fire the Research Center's Director, while others wanted to shut down the entire center. Instead, a small cross-functional team looked at the work being done and talked to customers (internal businesses) who did or could use the R&D Center. I did this in the days before "benchmarking" was a hot word, but we visited R&D Centers at a few other companies. The team's recommendation was to give the Corporate R&D Center to the two Divisions (i.e. a group of profit centers) that used it the most. When we presented to the Executive Committee, they asked about what other ideas the team had. They asked about what sort of return the company might realize, if the R&D unit just served two Divisions. And, to make a long story short, they really did not like the recommendation of not making any personnel changes and just having two Divisions serviced.

Now our young author's first reaction to this conundrum was that the Executive Committee was resistant to change, and he tried to sell the idea harder. Picture a sword coming up from the middle of the floor where our young colt could impale himself and bleed on the floor. He believed if the executive team saw enough blood and realized the degree of his emotional commitment to the team's recommendations, they would go along with this idea. The reality was no one really cared how much the young man bled. Fortunately for our young author, a couple of people on the Executive Committee liked him and called a halt to this painful process, inviting the young man for a chat after the meeting ended.

At that point the "Rule of 3" was explained. Not certain what is so magical about the number 3, but when teams come up with at least "3 Ways" to do something (solve a problem, capture an opportunity), somehow the third one seems to rise above the other two. The Rule of 3 sometimes results in a breakthrough solution on the third pass, provided the first two solutions are real alternatives, not just one idea we really want to do and two other throwaway ideas.

The first solution tends to address the primary problem that the organization is experiencing. If the team comes up with a second workable solution, it tends to be an extension or slight modification of the first idea. If they have the mettle to push forward and develop a third workable solution, it tends to be more of a breakthrough type idea, a different way of looking at the problem and developing a meaningful solution.

A project team working on a controversial project should never have just one major recommendation. There are several benefits to doing this. First, you have choices beyond simply falling on your sword if the first idea does not fly. More importantly, team members realize that there is more than one way to accomplish an objective. That automatically makes team members more receptive to questions from the key stakeholders. This is another way to create more space inside an organization.

The R&D team regrouped and looked at the problem again. We developed a second solution, but it was a slight variation of the first idea. The

Roles and Responsibilities of the R&D Executive Director were reviewed, and the team realized that the profit centers should have stepped up early in the game to provide direction to R&D personnel. It was a pretty passive relationship from a profit-center perspective (e.g. wait and see what these people come up with, and if it interests me, I may commit funds to it).

The third solution tried to address these issues. The final recommendation approved by the Executive Committee allowed R&D personnel to spend 15% of their time working on whatever technologies or ideas interested them (staying within budget guidelines). Eighty-five percent of all R&D work needed to be funded by a Profit Center at the start. Several new communication channels were provided to stimulate this. The R&D Center was also to focus on just the food side of Beatrice's businesses, not the manufacturing operations. The food divisions accounted for about 70% of Beatrice's overall revenue dollars.

The result of this activity was that 40% of all R&D projects now made it as far as a market test. Millions of dollars of new revenues resulted from the project, and staff turnover at the research and development facility decreased by over 25%.

Might this problem apply to you? When you (your team) do project work, do you push to devise alternative ways to solve the problem? There is almost always more than one way to do something. Force yourself (your team) to develop several pathways.

Problem or Process Issue

This can be an excellent way to change your perspective and make it more likely that gains from improvement activities will be sustained. Dr. Deming and several other famous people have stressed "that 85% to 95% of all productivity and performance problems result from an organization's processes, not the people who are working in the organization." When sharing this statistic during keynote presentations, people in the audience often object to such a high percentage. But always try to look at what is happening with a broader view, and think about the word "Process" with a capital "P."

For example, in the new product development process, does Design Engineering meaningfully interact with other departments and customers for the product, or is it designed with an internal view and tossed over the transom to the next group, with only minor interactions along the way? If an employee is having problems doing the work, is there an onboarding process? Is it effective? Did we effectively train the person for the work they are doing using training within industry (TWI) or some other effective protocol? Are the accountability processes fair and consistent?

Even though people *intellectually* understand there is an input–process–output relationship between the distinct steps of a process, leaders still often try to manage the individual process step(s) independently from the others. Senior leaders reinforce this thinking by challenging the owner of a part of the process to optimize a particular area of responsibility. *That view of the world makes it nearly impossible for the overall process to work in harmony.* You cannot optimize each process step in isolation; every step needs to work harmoniously with the other players to serve customers best.

Do improvements get made within a single silo (department), or are the effects of changing a process considered relative to internal suppliers and customers for the work being done? Problems with these activities occur due to poor interfacing in cross-functional process connections. Daily management activities at "elite" organizations always implement improvements with a process-centric perspective.

People tend to get pretty good at handling problems within their workgroup (department) because they are an immediate disruption to getting work done. Unfortunately, the normal way these issues often get addressed is to do a workaround. *Workarounds are the enemy of effective leadership practices.* We will come back to this thought in the Build Relationships chapter (Chapter 4) when we discuss things that destroy leadership's credibility.

Ensure you and the people on your team understand the overall process for the most critical cross-functional processes that are part of your daily life. Know what is important and seek to understand the major obstacles that inhibit that process's ability to create meaningful value. We revisit this thought in Chapter 3 "Define a Unifying Purpose" in the Unifying Purpose – Things to Think About section.

Better Leader Learning Cycle

At the very beginning of this chapter, we discussed the importance of taking time to reflect on learning to see things tomorrow that you do not see today. Reflection causes us to ask questions about our experiences. Let's practice becoming more effective leaders.

Your first step might be asking, "What did you learn from reading this chapter?" Write your response.

Consider some of the specific questions/thoughts discussed in this chapter:

1. Why do you want to make a change? Why is this important?
2. What actions must you take as a leader to create an environment for team members to feel safe in pointing out issues and problems?
3. What actions do you need to take to foster more critical thinking skills in all team members and to further engage your team members in solving problems?
   a. Problem identification/definition.
   b. Gathering meaningful information.
   c. Analyzing and evaluating alternative solutions.
   d. Reviewing/challenging assumptions.
   e. Communicating in a holistic and effective way.
4. How might you hold yourself accountable for the changes you wish to make?
5. How do you plan to recognize and reinforce progress?
6. How might you deal with obstacles that inhibit moving forward?

Your responses to the above questions will strengthen as you gain new insights. Start by focusing on the first four questions. Actions to address questions 5 and 6 will become more apparent as you practice with experimentation and reflect on the results.

Reflection sets the stage for learning. As we change our perspective through new learning, we gain new insights that increase our abilities to lead more effectively. Get some early input from a trusted/respected person. I encourage you to converse with at least one other person and share your thinking.

- **Potential Actions**: What are two actions you can execute to learn something new about the way you lead, your area of responsibility, your team, or your organization?
- **Practice**: What is your practice plan for those actions? When will you start (date/time)? How will you hold yourself accountable to practice these actions daily or weekly? The last chapter in this book (Chapter 6) lays out a way to practice developing new habits and new behaviors.

- **Evaluate**: How might you validate that any new actions taken are more positive than negative regarding their impact on your team, your department, and your peers?
- **Next Steps**: What do you plan to do next based on what you just learned? Your responsibility as a leader is to influence the people with whom you interface. The most essential part of that influence is simply doing or getting done what they have tasked you to do. But more is needed for great leadership. Outstanding leaders uplift the people around them. If you effectively do this, the results can genuinely become transformative. It is cool when it works, and you will be a better person because of it!

The real voyage of discovery consists not in making new landscapes but having new eyes.

Marcel Proust

Notes

1. A version of this story was also shared on Mark Graban's blog – https://www.leanblog.org/2018/05/how-200-jobs-were-saved-by-engaging-employees-in-continuous-improvement/.
2. This quote has also been attributed to Mark Twain and Will Rogers.
3. Avid Reader Press / Simon & Schuster; Illustrated edition (September 19, 2017).
4. Alan Deutschman, "Change or Die." *Fast Company*, May 1, 2005; https://www.fastcompany.com/52717/change-or-die.
5. John Kotter, *Leading Change*, Harvard Business Review Press, 1996.
6. Mike Rother, Keynote at *AME Annual Conference*, October 2018.
7. Thomas J. Peters and Robert H. Waterman, Jr., *In Search of Excellence*, Harper & Row, 1982.
8. John Colby, *AME Annual Conference*, 2022.
9. Carol Dweck, *Mindset the New Psychology of Success*, Ballantine Books, 2016.
10. Richard Tanner Pascale and Jerry Sternin, "Your Company's Secret Change Agents." *Harvard Business Review*, May, 2005.
11. Stephen Covey, *7 Habits of Highly Effective People*, Simon & Schuster, 1989.

Write Your Personal Notes on Reflections Here

Chapter 3

Define a Unifying Purpose

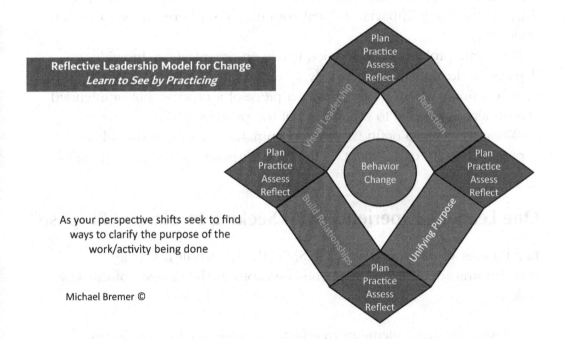

Reflective Leadership Model for Change
Learn to See by Practicing

Plan
Practice
Assess
Reflect

Visual Leadership

Reflection

Plan
Practice
Assess
Reflect

Behavior
Change

Plan
Practice
Assess
Reflect

As your perspective shifts seek to find
ways to clarify the purpose of the
work/activity being done

Build Relationships

Unifying Purpose

Plan
Practice
Assess
Reflect

Michael Bremer ©

What Is Your "Purpose"? Does It Drive the Right Behavior?

Having a meaningful purpose, where we hold ourselves accountable to do
what is right for the overall organization, is something that (once again)
seems like it should be so simple to do. Yet, it is a challenge.

Many books and people talk about the importance of "purpose." Just
because you have a purpose does not mean it's necessarily beneficial. Is it

DOI: 10.4324/9781003495284-3

the right purpose? Let's be clear: this is not about pleasant words written on a piece of paper. That's easy to do and, once published, easy to ignore on a day-to-day basis because we don't hold ourselves accountable for living that purpose. There is a real purpose for your work and how you lead, but when it's not clearly defined and meaningful, it results in inconsistent leadership behaviors and confusion for team members. It's important to discover your real purpose.

For years public companies said their focus was "shareholder value." But a singularity of focus can have unintended consequences. A narrow focus on profitability caused many organizations to inappropriately outsource parts of their value-adding activities to obtain lower labor costs. This sometimes resulted in a loss of expertise for future new product development, and total cost considerations were rarely considered for supply chain disruptions or the impacts on local communities where the organization operated.

The same problems can happen inside an organization. Individual department leaders might narrowly focus on the performance metrics for their functions, where they optimize a piece of a process, but unintended consequences happen to other parts of the process.

When we seek to define a more meaningful purpose, it should be approached as an experiment. Is the new purpose the "right" purpose?

One Leader's Experience with Seeking to Clarify "Purpose"

Eric Pope is VP of Operations at U.S. Synthetics in Salt Lake City. Paraphrasing some of his comments, he captures the essence of this key action:

> There are many elements to effective leadership, but few are as important as giving people a unified and meaningful reason, "a purpose" for change. This reason must be compelling enough that people are willing to subject themselves to the anxieties and struggles of change. It must be motivating enough to create a willingness to try something new. It must feel safe enough to broadcast their work's problems and shortcomings and concede that there has to be a better way.

What is the purpose of your business? How do you define success? Most people answer this question with business language (e.g. to be the best at ... or to be the industry leader at ... and they would finish their story of success and purpose by describing great business results, market share, sales, profit, growth, etc.).

But what if I asked about your personal purpose in life? Some successful people will say it's all about making money. But when you probe more deeply, you learn that money is simply a metric they use to talk about their ability to influence others or, in some instances, their desire to control others to do their will. And if you continue to ask, "Why is that important?" ultimately, they typically want to have an impact. They want to leave some mark on the world that provides evidence of a meaningful life.

How would you describe success in your own life? Whenever I've gone deep with this conversation, people talk about personal success in terms of their family, the strength of their relationships, having fun with others, the ability to make an impact in the lives of others, and being a force for good in the community. I have found that wherever I go in the world, no matter the demographic, people will have the same general purpose and vision of success. It may be scaled differently. A senior executive at Apple might not use the exact words as a community organizer in Togo. But their intent and desire to leave something positive behind are similar.

Now, take those two views of success, business, and life and ask which is more motivating. Which one will people get passionate about doing? Which one will carry them when times are tough? What purpose will foster alignment to work collaboratively and cooperatively with the others who surround us?

Our job as leaders and builders of culture is to establish a compelling and shared purpose. That shared purpose already exists in the minds of every employee. They already work tirelessly for it and are committed to and passionate about it. And it's not your "be the best in the world at ... vision." It is about lives. It is the reason why we all go to work. It is the reason why the business exists; it is to improve lives. Leaders must Connect Personal Purpose to Business Purpose!

Eric's comments apply to the organization as a whole and inside to each functional group. They apply to individuals and to our role as a leader. What are we seeking to accomplish holistically (for our department/team)? And in the case of improvement, what is our purpose for change? A meaningful and desirable purpose can instill passion in people. This chapter has examples of company and department/team purposes. As Eric noted, a real purpose connects to your personal purpose in life.

Unifying Purpose – Things to Think About

Discover a Meaningful Purpose and Share It!

Most leaders we have observed over the years do not write down a personal purpose for their work, why they are leading, or what they hope to accomplish as a leader. Similarly, when organizations use necessary lean improvement tools (e.g. Leader's Standard Work, Gemba Walks, etc.) that they hope will help change their culture, they rarely write down a clear purpose for using these tools.

Improvement can still happen without a purpose, but when you write down:

- What do you expect to accomplish as a leader?
- What is the reason for the existence of your team or department?
- Why are you using this particular improvement tool?

You are then in a much better position to hold yourself accountable and periodically assess how effectively your purpose is being fulfilled or if that improvement tool is working effectively. Is the purpose being accomplished? It is amazing how many people ignore this simple but powerful step.

Purpose, as I am using the word in your role as a leader, should touch on three things:

1. **People:** What are you doing as a leader to develop your people?
2. **Process:** What are you doing as a leader to improve the performance of your processes?
3. **Performance:** What are you doing as a leader to improve the performance of the overall organization (not just your team)?

Purpose differs from goals. A "goal" is a quantifiable target. The purpose is more focused on why this goal is an important thing to do. You will note that I'm focused on "improvement." *Clarity of purpose* helps to guide what we should be improving.

Relative to leveraging your leadership role, if your purpose does not touch at least two of the above 3Ps, it is narrowly defined and does not take a holistic view of what you should be doing.

Can you effectively explain why you are doing this activity as a leader? Why are you using this tool or technique? What role are you serving as a leader driving/guiding/coaching improvement? What is the reason "why" you wish to XXX? Clarity can significantly help you improve the way you lead. *It's worth spending time reflecting on the purpose of your role as a leader, the work you (and your team) do, and the protocols you use daily to get work done.*

At a department level, a unifying purpose needs to shift away from the department's activities and instead focus on the (usually internal) customers the department serves. Consider a hospital blood testing lab. Its activity is to conduct analysis and send a report with the results. But its purpose differs. The lab's primary purpose is to provide physicians with the information they need to accurately diagnose a patient's health. A compelling purpose statement for the hospital lab should look directly at the customer for the work being done. You are much more likely to arouse people's passion for the work they do if they understand *and support* an external vs. internal (activity) view of their work. The group's leader needs to "walk the talk" for this type of purpose if it is to be meaningful.

A meaningful "purpose statement" seems to be a vastly underutilized component of the improvement toolkit. All too often, people use improvement tools, assuming the purpose is obvious: "implement an improvement or let's do leader's standard work." It also happens inside departments: "we process insurance claims, or we make this product." However, those statements are activity-focused and do not provide a meaningful reason why this is important. Without a clear purpose (i.e. why are we doing this), it is more difficult to challenge the effectiveness of the activities being done, as well as the tool being used and the action just taken.

A question I always ask when I take part in a site visit to a company applying for AME's Excellence Award is, "Why are you using this improvement tool (e.g. leader's standard work)?" It is incredible how few people can answer that question with a simple statement. Typically, they use the tool and assume it will be helpful. But without clarity of purpose, they are less likely to reflect on how they use that tool and what actions they might take (behavior changes) to use it more effectively. If the tool does not work as expected, they stop doing it. *So, a learning opportunity gets lost.*

Organizational Structures Inhibit "Doing the Right Thing"

Most organizations operate with departmental silos that look like the pyramid diagram in Figure 3.1, with the CEO at the top and functional workgroups, departments, and business units working underneath.

We have an exercise we have played in our workshops for over 25 years where teams operate in a typical pyramid structure. It is a powerful learning experience, and it fits with clarity of purpose or lack thereof. *If you want to run this exercise, contact me and I'll send you a guide.*

Here is an overview. The instructions given to people at the beginning of the exercise are ambiguous and they attempt to interpret the "real meaning" of the stated goal for the exercise which is: "Win as much as you can."

Participants are divided into four divisions of the ABCD Company. At the beginning of the exercise, they may not talk to people in the other silos. During the exercise, periodic "bonus" rounds further influence people's behaviors. Later, when they are allowed to speak with people in the other silos, the agreements they make are not always crystal clear or kept.

During the exercise, they need to trust the other teams (silos) to do the right thing. Given the way the training is structured, most teams seek to maximize the score for their silo at the expense of the other players.

Questions we ask when debriefing the exercise include:

- What was our purpose in this exercise?
- How effectively did we stop/look/listen?
- How hard is it to "do the right thing" when you feel others are not playing fairly?

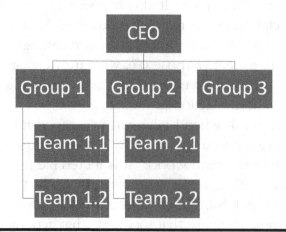

Figure 3.1   Organization chart

- Does that justify your actions?
- How much is this exercise like reality?
- What would you (we) need to do differently to obtain a better outcome?

The pressure to not do the right thing is strong, and unfortunately, the behaviors that are experienced during the exercise mirror what happens in the real world when the purpose is not clearly defined and when the purpose is internally focused on what we do in our silo, our department, our team, rather than how we impact others. This is why it requires courage for an individual leader to step up and do the right thing for the organization. Visual Leadership is discussed in Chapter 5 as a lever that can help the right thing to happen, even at an individual department or project team level.

Terri Kelly, the former CEO of W.L. Gore, captured the essence of establishing a unifying purpose in a video I saw a number of years ago. She said,

> Create the right environment where collaboration happens naturally – where people actually want to work together, where they actually like to be part of something greater than just the individual contribution. And if you get that part right, all the other pieces fall in place. That way of operating has allowed us to create this great innovation cycle within Gore.[1]

Creating this environment is a challenge! There is ample evidence to show most organizations do not do this well, with employee engagement survey results being one piece that shows there is a problem. What can you do as a leader to make certain your team has a meaningful and holistic understanding of the purpose for doing their work?

Clarify Purpose

A Unifying Purpose can help to break down the walls intentionally or perhaps unintentionally built between departmental silos. I have seen the following diagram multiple times over the last 30 years (Figure 3.2).

Unfortunately, most organizations still manage by functional unit or department. There is a much greater "intellectual awareness" of cross-functional processes today vs. 30 years ago, and indeed customers get served in this fashion. But traditional organizations still primarily measure the performance of individual departmental silos. While no one says this explicitly,

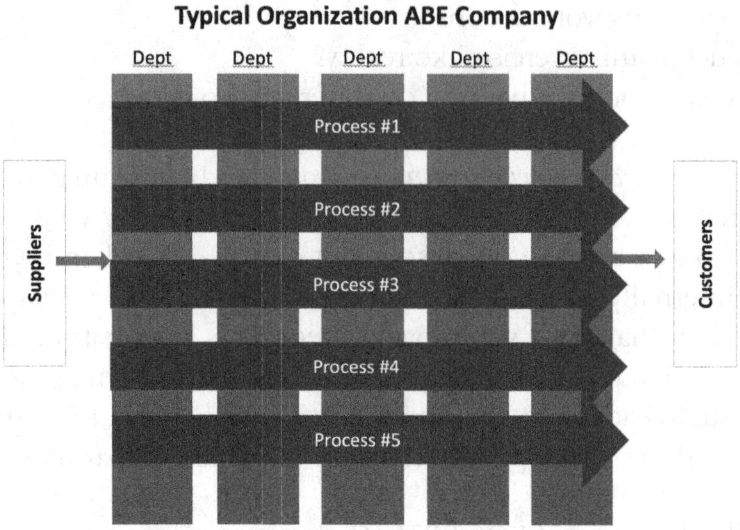

Typical Organization ABE Company

Dept   Dept   Dept   Dept   Dept

Process #1

Process #2

Process #3

Process #4

Process #5

Suppliers

Customers

Figure 3.2   Organization silos

the mental model is, "If every department maximized its performance, we would do a great job of serving our customers." Traditional performance metrics reinforce this view. Beyond on-time delivery and quality metrics, which are somewhat cross-functional, the rest of the metrics typically focus on the performance of the individual silos.

You cannot effectively serve customers by maximizing all of the component pieces. The silos need to work in harmony. Maximizing what is happening in some of the silos can make it more difficult to serve customers. Think of the disconnects you have probably seen in your life between Sales and Manufacturing or Engineering (design) and manufacturability of the design. Short-term goals vs. long-term strategies. It's better today than it was 30 years ago; there is more cooperation, but we are still far short of what could be done.

Make Sure Your Purpose Is Both Real and Meaningful

People in the defense industry like to say, "We exist to help the warfighter," and ultimately, that may be true, but that is not the primary purpose of most departments. Focusing on something that is far removed from the work they are doing makes it easy to justify whatever they want to be doing within their organizational silo. It assumes that their department, team, or group is the best source of creating value for the warfighter.

That view neglects the fact that your silo needs to coordinate/collaborate with other departments, teams, or groups that also feel they are the primary source of value. To create something meaningful and useable to "the warfighter," the internal silos need to collaborate and cooperate. For example, aircraft software developers might be doing something related to tracking targets, but how does their software impact the ability of maintenance personnel in the field to do their work? If the plane is stuck on the tarmac because it's difficult to maintain, that isn't doing the warfighter or anyone else much good.

This requires leaders to have a deep understanding of the overall cross-functional processes used to create value for the warfighter or whatever customers your organization serves. Unfortunately, the metrics used by many organizations focus on optimizing value for each silo within the overall process. Cross-functional processes do not get optimized by maximizing each component part! (Figure 3.3). That is one of the neat strengths of the "Lean" movement when it focuses on flow and finding waste that inhibits the creation of value for customers.

Figure 3.3 Which way should I go?

A second example: Assume an organization manufactures a customized highly engineered product. This is a story I'll expand upon in the Visual Leadership chapter (Chapter 5) with One Leader's Experience – the Request for Proposal story. Of course, engineering works to create a great design; they seek to win the request for proposal as their organization bids against other companies. So, those activities are part of their purpose. But does it stop there? Is that the only important purpose for a Request for Proposal (RFP) process? Engineering receives a request for a proposal from the Sales Department. What is Sale's purpose for their part of the process? The individual salesperson is hoping to get a commission on a successful RFP bid. Is that the purpose? Or, is it something else? The true purpose often goes undefined. Hence, design engineering might be working on an RFP where the outside customer has no real intention of considering this bid. The purpose of a Request for Proposal process needs to take into account all the key players in the process. And metrics the organization is using should help it learn how to make the overall process work more effectively. Read the story in Chapter 5 as it has an elegant and very successful resolution (Figure 3.3).

Improvement at P&G at a Corporate Level

In "Leading Change: An Interview with CEO of P&G" Alan Lafley *(now retired)* said, "It's important to raise aspirations, but I agree with Lou Gerstner's position *(former CEO of IBM):* strategic visions can be a distraction." Lafley also seconds Lawrence Bossidy's belief that companies should aim to achieve great execution, but insists that the real challenge is to "unpack" this idea.

> Bossidy's right – in the end, it's about executing with excellence. But you can exhort all you want about excellent execution; you're not going to get it unless you have disciplined strategic choices, a structure that supports the strategy, systems that enable large organizations to work and execute together, a winning culture, and leadership that's inspirational. If you have all that, you'll get excellent execution.[2]

It's important to unclutter the thinking of team members/associates so that they can focus on the critical business of problem-solving. People have so many things going on in the execution of their daily business activities that they don't always take the time to stop, think, and understand the overall context. When they can see what it all really means (clarity of purpose) and how it fits with what the organization is strategically trying to accomplish, they can do a much better job of finding improvement opportunities and implementing change. Lafley said, "I want them to use the same basic model and the same discipline to make the right choices for, say, the Philippines," *where P&G has a half-billion-dollar business – a sizable operation but only 1 percent of the whole.* "I want the manager there to think very consciously about what kind of culture is going to be a winner, what kind of capabilities are needed, and so on."[3]

Whereas this purpose statement from a training department – "Our mission is to ensure that a well-informed, prepared and motivated employee population can effectively perform company processes and practices." This is OK; I've certainly seen worse. But there are many causes outside of a "training department" that impact well-informed and motivated employees. Contrast the training department example with this one from an Information Systems group. Their revised purpose stated, *"Our focus is to deliver information that helps the business make decisions by providing value-added solutions"* [emphasis added]. Both examples are nice words. I would want to know how they are measuring their execution effectiveness. They could possibly measure how often they are called in early for discussions, lead time to implement a solution, do the IT solutions work, does the training get used in a meaningful way, are their internal customers growing their businesses?

Whether you call it Purpose, an Aspiration, or the Reason Why isn't important. What is important is for people to have an accurate, concrete understanding of the context for the work they do? How does "my work" fit with what the organization, the department, or the team is trying to accomplish? Clarity and meaningfulness of "purpose" are powerful tools for driving appropriate behaviors and empowering people to make better decisions. What is your reason why?

Move Beyond the Obvious

Our job as a leader is to connect our work to a meaningful purpose – that is bigger than the work task. To do so, we must ask ourselves how our business, department, or team enables, helps, or facilitates that bigger picture. It does not matter if we look at the business, a department, or a team. The Purpose is something beyond the border of the activity. A business is expected to be profitable; a factory is expected to make a quality product and deliver it on time, and a governmental unit is expected to deliver services in a timely way. Organizations are expected to do those things; there is nothing unique about being profitable or knowing how to do the work task. If you want a motivated team or workforce, you must reach beyond the obvious.

When I was first involved in the world of performance improvement, the expression people used was "Walk the Talk." What that expression meant was that it is easy to say the words, but it is hard work to hold yourself accountable to *living your life that way*. The pressures to focus on the short term are considerable. It will take some courage to change your behavior and your focus.

A meaningful connection links the business and work activities inside the organization to a variety of possible purposes:

- Provide goods and/or services to make the world a better place *(easy words to say; the challenge is understanding how what you do makes that happen. What is the meaningful metric?)*.
- Provide jobs to employees so they can support their families and their life's aspirations, even if the current job is simply a stepping stone along the way.
- Do no harm to the environment.
- Give life to communities where the organization operates; this facilitates bringing new talent to the organization because the community is a desirable place to live.
- "I believe the business exists to improve lives, not to make people's lives smaller," says Richard Sheridan, CEO at Menlo Innovations. He feels work should bring "joy" to people's lives, and he leads his business aligned with that value.[4]

We must formally make this connection, declare it, message it, and live it.

Power of Purpose at an Individual Level

Nick Craig co-authored *From Purpose to Impact* with Scott Snook. In their book, they talk about purpose as a life statement. Nick said,

If you don't know it, you can't fully live it. And if you aren't living it, you can't lead from it … When you get clarity of purpose, you see the world through a unique filter, and this gives you the opportunity to be more creative and innovative about how you lead your life. It creates "meaning" from events and actions that shape your impact on the world.[5]

I will not go as deep here talking about purpose, but if you are interested, I encourage you to read Nick and Scott's book. In a *Forbes* magazine interview, Nick was distinguishing between causes and purposes. They are not the same. Your cause may be to end world hunger or to stop discriminatory practices. But that is not your purpose. He said,

As you think about your role in the world and the difference you will make, ask, "What is the unique gift that I bring to what I do? What would people miss most if somebody with equal skills replaced me?" In expressing these aspects of yourself at work and beyond, it is entirely possible to end up saving a small corner of the world along the way.[6]

Your personal purpose is about something that is very important to you and is something you would want to do in any job you hold, not just your current position. *Giving some thought to Nick's question about what people would miss if you were not there is a good reflection question.*

Passion

Many (if not most) of us feel our lives should have meaning. Yet when you look at employee engagement scores, clearly most of us are not passionate about the work we do. You will not be passionate about everything, but make sure you are passionate about what is important. There are at least four aspects to this:

1. **Your Family:** includes prior, current, and generations to follow.
2. **Your Work:** includes your team and your company.
3. **Your Community:** includes the people who live near you, people affected by what you do, and your faith-based practices relative to elevating others.

4. **Your Self:** this comes last on the list for a reason. If you focus on others first, your life is likely to have much more meaning. *No matter what your faith, when you die, your possessions do not go with you. Life is about relationships.*

Don't rationalize your way to the land of mediocrity. Sometimes we get a little misguided, and it is surprisingly easy to stray from the path. Sometimes we get trapped by the decisions we allow others to make for us. If we let others decide how we live our life, that is a choice we make! We have alternatives, but you cannot be passive; you must act!

If you believe you are passionate but the results in your life fail to yield what you hoped to accomplish in your relationships, then perhaps you are passionate about the wrong things or perhaps you need to overcome some fear. You probably need to make some changes if you wish to be more successful.

Richard Sheridan, the CEO of Menlo Innovations, a software development company, said,

> After one of my talks, someone sent me a link to a Simon Sinek video saying, "You naturally do this." But when I listened to Simon's discussion about "why" I realized I might touch on why, but I don't start there. Nor do I use "WHY" effectively. I then probed our mission. One of our goals is to reduce the "suffering" of people who use software. It's important! But when I thought about it, that was not our key WHY. Probing further I thought, "Our goal is to return Joy to the development of software. That is our key why."[7]

This is a good example of reflection and refining your purpose statement; it should link to "why" this is important.

As a leader, you need to raise people's aspirations. But you don't want to get all caught up in the vision and mission game. In the 1980s and 1990s there was a wave of companies drafting Vision, Values, and Mission type statements. But very often over time, those statements lacked something important … they lacked accountability. Leaders did not "walk the talk." Purpose without accountability ends up being "nice words on a piece of paper." If you define a meaningful purpose, you can build in accountability by making it part of what we later discuss in Chapter 5, Visual Leadership.

Autoliv's U.S. operations are located in the State of Utah. The manufacturing operations are not near their customers' manufacturing operations around the world. Part of the "purpose" discussion between leadership and associates at Autoliv focuses on the fact that if they (leaders and work associates) wish to keep high-paying jobs in the State of Utah, then they need to figure out ways to manage their costs, improve their quality, and service their customers more efficiently and effectively than any other manufacturer of airbags around the world. This is a key part of the communication messages at Autoliv. Importantly, leadership does not "tell" people they need to do this; instead, they create an environment where people are not afraid to share their thoughts and are empowered to take action.

Spend some time thinking about the purpose of your work, your team's work, and your department's work. The Actions to Practice section below describes a few ideas you might use to move in this direction. Take an expansive view and think beyond your department's (or silo's) responsibilities. What is the role your team is playing in the process? As a leader, when you and your team members look beyond your silo, you can begin to develop new metrics and targets that positively influence cross-functional cooperation.

Find a few ways to hold yourself accountable to the agreed-upon purpose statement. The Visual Leadership chapter (Chapter 5) shows several ways to practice doing this. The proof comes down to the way leaders behave during "crunch time." Does the leader cave and ignore the agreed-upon purpose and do what is most convenient at that point in time, or do we focus on the long-term goals of what we are trying to accomplish and do the right thing?

I'm not saying you should *never* do a "work around," but what I am saying is don't automatically do a work around. When a problem comes up, first try to understand from a process perspective why the process did not work. Often, you will not discover the root cause of the issue right away, but it is critical that you take a more holistic perspective rather than jumping into "working around the problem." Do exceptions exist? Of course, if a tank is leaking, plug the leak. But don't stop there.

Actions to Practice – Define a Unifying Purpose

Steps to Create a Meaningful Purpose

As noted earlier, when talking about "purpose" in this book, it is relative to effective improvement practices and how leaders interact with people; we are not talking about nice words written on a piece of paper. No doubt, you most likely became a leader because you are reasonably good at getting things done and solving problems. If you are going to make a shift, as we later suggest in the Build Relationships chapter (Chapter 4), to lead with more humility and elevate your teammates, peers, and others, then unity of purpose can help make that happen.

1. **Discover What is Meaningful**: A purpose already exists for your team and for the way you lead; it might not be formally recognized, and day-to-day behaviors might not be consistent with the intended purpose. Ask questions to the people on your team to discover your "real" purpose. What have they done that felt inspirational? When did they feel they connected and did something important for your direct customers or the external customers for the work you do? What happened in the past that made people feel the work they did was meaningful and appreciated by your customers (who may be internal to the organization)? What do you do as a leader that positively reinforces the overall purpose; what do you do that might inhibit progress?

2. **Synthesize the Ideas Gathered**: Float them by the members of your team and, if possible, with the customers you serve. Try to get something that is crisp and meaningful, and you, as the leader, and the people on your team, can be held accountable for doing. Earlier in this chapter, we referenced Richard Sheridan at Menlo Innovations. He felt his organization had a meaningful purpose, "reduced the "suffering" of people who use software."[8] This is important and it is certainly noble, but it is not measurable or even actionable in a visual way day to day. When he and the team further explored their purpose and expressed the thought, "Our goal is to return Joy to the development of software." While they did not talk about external customers in that statement, they were a part of the focus. Software developers regularly talked to the external customers for whom they were doing development. Assuming Menlo's leadership holds itself accountable for doing this day to day, it becomes quite actionable and can drive behavior change. So as the questions get asked, seek to go deeper than a high-level (but noble) platitude and find something that will assist in driving/guiding behavior change.

If you have reasonable clarity of purpose and, for some reason, the purpose gets violated, seek to understand why it happened as quickly as possible right after the incident. There are learning opportunities that may get missed if a holistic perspective is not taken right at the outset.

Become a "purpose-driven leader" and you will be amazed at the profound behavior changes that will emerge from the people around you. People generally want to do the right thing; they want to do something meaningful. Your responsibility as a leader is to nurture those behaviors, to foster more critical thinking skills, and build the self-esteem and confidence of your team members.

Probe to Better Understand and Define the Activities You Do

For a department within an insurance company, "We process claims" is a task statement whose that purpose is obvious and not very exciting. However, if the purpose was something more along the lines of, "Our department works in a way to seek a fair resolution for our customers' claims in a timely way" is a higher-level purpose that might be somewhat inspirational. Autoliv manufactures airbags for the automotive and trucking industry. "Making a high-quality airbag" is an obvious purpose. But figuring out "How to make airbags in a less expensive and faster way in order to keep high paying jobs in Utah" is a higher level and requires a more critical thought process.

Just saying these words isn't sufficient, but if the words used are supported by leadership actions and accountabilities, then the purpose statement can be very valuable.

You can learn more about your (leader, team, department, or improvement tool) "purpose" by:

■ Asking, "Why? We are doing this at least 3 times – why is this important?"
■ Take a walk and see how your customer uses your outputs (observe).
■ Talk to your customer's customer!
■ Take the metrics you use and share them with your customer (often internal), and learn how your measures impact the work they do.
■ Check how your goals and metrics measure performance against the purpose. Are they internally or externally focused? Do they hold you, your team, your department accountable for something meaningful, or do they merely focus on activity counts (beware of the activity trap)?

Consider your direct customer(s). What is the most important thing you would like them to say about the services/products your work group provides? If asked, what do you think your customers would actually say? Write it down before you have the conversation. How does what they said compare to what you wrote?

Note: The SIPOC tool in the Reflection chapter (Chapter 2) can also be used here.

Get Out of Your Office and Go to the Gemba

Gemba is a Japanese word. It means going to "the real place" (go see where the actual work is done and value is created). Gemba Walks represent a potential learning and development opportunity both for the walkers and the people being visited during the walk. The amount of learning, discovery, and associate development that takes place largely depends on how effectively the walk gets done.

Gemba Walks offer a way for the walkers to change their perspective on how they see, understand, and manage their organization. Gemba Walks are about helping everyone in the organization to more clearly see and understand the whole process as a first step in identifying ways to improve it.

Doing a Gemba Walk is more than "management by walking around." An effective Gemba Walk has a clearly defined purpose. The reason a front-line supervisor does a walk will differ from a plant manager, a director of a software development team, or a visit by an outside executive. Each of those can do an effective walk, but the purpose for doing it should be clearly defined and periodically assessed for how well it is fulfilling its purpose.

When I'm visiting a site applying for AME's Excellence Award, I'm trying to understand how value flows through the organization and their improvement protocols. My purpose would include (Figure 3.4):

A Gemba Walk to *Go See, Ask Why, and Show Respect* is a key way to engage people more actively in performance improvement activities. There are three key reasons for doing a Gemba Walk:

Mike's Purpose for Initial Walk

What does value flow look like?
- Review key priorities
- How managed day-to-day
- See key cross-functional interface points
- What issues addressed and how done
- Get feel for employee engagement - look for smiles, eye contact & comfort when asked a question

Figure 3.4   Mike's purpose for a Gemba Walk

1. **Clarify Purpose**: Gemba Walks provide a wonderful opportunity to learn if people inside the organization have a deep understanding of "why" they are doing their work activities.
2. **Gain Better Process Understanding**: Leaders can see, with their own eyes, how effectively work activities within or between departments align with what the organization is trying to accomplish. *If your senior leaders are not doing walks, perhaps you should try doing a cross-functional walk with your peers to better understand the creation of value inside your business.*
3. **Engaging People**: The walks provide an opportunity for leaders to discover barriers that inhibit people's ability to do great work.

Mike Hoseus, former Assistant General Manager at Toyota Motor Engineering and Manufacturing, said, *"For us, the Gemba Walk was a way to live out the 'Servant Leadership' principle. A big part of our focus was building a relationship of mutual trust and respect on our walks"*[9] [emphasis added]. A first-line supervisor might have multiple purposes for different walks done during the week (Figure 3.5):

The *How to Do a Gemba Walk* book goes into much more detail on this powerful practice if you have an interest in learning more. You can also download a set of free slides from LinkedIn's Slide Share if you search for the book title and my last name. The slides provide a reasonable overview of the content within the book.

First line leader's walk

- Tuesday 8am – Safety Walk (ergonomics/or reaching, looking for things, safe handling mat'l....
- Wednesday 8am – Progress on existing action items
- Thursday 8am – Standard Work Observation (share Autoliv story)
- Friday 8am – Seek improvement thoughts on reaching next target condition

How might this improve the quality of a walk?
What else could make them better?

Figure 3.5 Example: Supervisor's purpose for a Gemba Walk

How Meaningful Is the Work Done in Your Department?

If you were to ask the people in your department, work group, or team why it exists, what would they say? Write it down. Next, ask people from your group to write the purpose as they understand it. Do not share your copy! How do they compare? Have a discussion to reconcile differences in the multiple views.

Consider using this simple survey to get feedback on the meaningfulness of work done within your group (Figure 3.6):

When you reconcile multiple views, consider talking to people from the departments you serve. Get their thoughts on how your work helps their group to serve their customers. How does the output(s) from your work group, make a difference to customers or to the rest of the organization? How can you validate the above scores with evidence? If you have changes

How Meaningful Is Work?	
For your Department, Work Group or Team	Scoring
	7 very high - 1 very low
Majority of people are passionately engaged in making or doing the work, they care about their direct customers (may be an internal work group)	
Our work is supported by clear and meaningful performance targets (not activity counts)	
People **inside** our work group take 'pride' in the group's outputs	
Outputs from work done by our group are meaningful to the people we serve; they respect and appreciate the work we do	
We use publicly posted, visual feedback to foster accountability and collaboration; we know if we are winning or losing in trying to progress toward key targets	
We know what obstacles inhibit success and take action to address them; gains are sustained and replicated elsewhere when appropriate	
Total Score (maximum is 42 points)	

Figure 3.6 How meaningful is work survey

you want to make, how can you run an experiment to see if the changes implemented are positive?

The goal here is not to see how high you can score. The goal is to use this instrument to foster a dialog and to change your perspective on improvement opportunities. Have your team members reflect on how their work does/does not fit with the mission. Then, use goals and metrics to better align toward your clarified purpose. Metrics should provide rapid, meaningful feedback. Coach your team members and create an environment that encourages experimentation, rapid learning, and accountability.

Better Leader Learning Cycle

It is critical to help people understand the purpose from an external perspective. This applies from the top to the bottom of the organization. The organization also needs clear strategic direction and a meaningful business model. But if you are somewhere in the middle of the organization and the overall purpose lacks clarity, that is not a good excuse for your team to lack clarity for its purpose. *Figure it out!*

Realize you are working toward a greater good, and it takes time to learn how to improve more effectively and to determine a unifying purpose. Figuring it out requires a willingness to experiment. Experimentation requires allowing failures to happen. Some people say, "Celebrate failure!" I think that is misguided and see no reason to celebrate if we fail. Failure is fine, so long as we learn from it. What we don't want is to be afraid to fail or to admit failure. And we don't want to miss the learning opportunities.

What did you learn from this chapter? Are you willing to take the simple assessment in the Practical Tools section and use it to start a dialog about "purpose" for your work team? Can you use the questions we shared in this chapter to get started?

- Asking, "Why? We are doing this at least 3 times – why is this important?"
- Take a walk and see how your customer uses your outputs (observe).
- Talk to your direct customer's customer.
- Take the metrics you use and share them with your customer, learn how your measures impact the work they do.
- Check alignment of your goals and performance metrics against your purpose statement.

- **Potential Actions**: What are the two actions you can execute in the next 30 days to better define the purpose for your area of responsibility, your team, and your organization?
- **Practice**: What is your practice plan for those actions? When will you start (date/time)? How will you hold yourself accountable to doing it?
- **Evaluate**: How might you validate that any new actions taken are more positive than negative in terms of impacting your team, your department, and with your peers?
- **Next Steps**: What do you plan to do next as a result of what you just learned?

Clarity of purpose for your role as a leader, for the work done by your team and/or department, can help to prioritize which actions are most important. Additionally, it can help (along with effective Visual Leadership practices) to drive more collaboration and cooperation across departmental silos.

Clarifying your purpose and finding ways to hold yourself and your team accountable for behaving in an appropriate fashion can go a long way towards generating more passion and engagement in your work.

Notes

1. https://sloanreview.mit.edu/article/creating-a-culture-of-innovation/ February 13, 2009.
2. Rajat Gupta and Jim Wendler, *The McKinsey Quarterly*, Web-exclusive, July 2005.
3. Ibid.
4. Richard Sheridan, keynote, AME Annual Conference, 11/01/2018.
5. https://www.forbes.com/sites/forbescoachescouncil/2018/07/05/.your-cause-is-not-your-purpose-heres-how-to-tell-them-apart/.
6. https://www.forbes.com/sites/forbescoachescouncil/2018/07/05/.your-cause-is-not-your-purpose-heres-how-to-tell-them-apart/.
7. Association of Manufacturing Excellence (AME), "Keynote Presentation" October 2022.
8. Association of Manufacturing Excellence (AME), "Keynote Presentation" October 2022.
9. Jeffrey Liker and Mike Hoseus, *Culture the Heart and Soul of the Toyota Way*, McGraw-Hill, 2008.

Write Your Personal Notes on Purpose Here

Chapter 4

Elevate the People Around You – Build Relationships

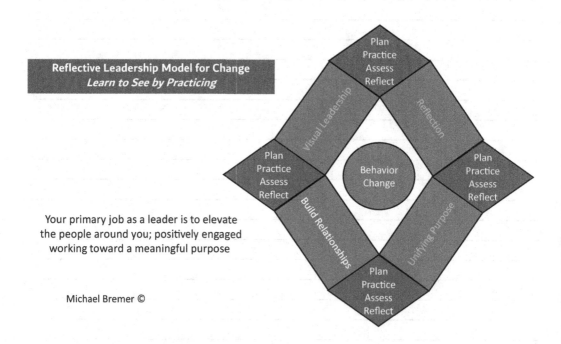

Reflective Leadership Model for Change
Learn to See by Practicing

Your primary job as a leader is to elevate the people around you; positively engaged working toward a meaningful purpose

Michael Bremer ©

People Desire Meaning in Their Life – I Matter!

> I want the fact that I existed to mean something. And I'd rather make a mistake, than do nothing.[1]

Those are two statements from a singer/songwriter who told wonderful stories in his music and died much too young – Harry Chapin. If you are not familiar with him, please listen to some of his songs; they are timeless.

DOI: 10.4324/9781003495284-4

Most people desire to be part of something meaningful, to be part of something greater than themselves. It can come from family, a community, a nonprofit group, or part of the work we do. Ideally, several of those life experiences give meaning to a person's life. An insufficient number of leaders take responsibility to create an environment that helps to give meaning; all too often we fail to help people gain self-esteem and confidence. It's easy to complain and tear things down; it requires energy and effort to be a builder. Maya Angelou once said, *"People will forget what you said, people will forget what you did, but they will never forget how you made them feel"*[2] [emphasis added]. And that can be positive or negative. We can all learn how to do this more effectively if we are willing to practice getting better at building people and learn to see the invisible that surrounds our daily lives.

You don't become whole as a person until you become part of something bigger than yourself. People are willing to become much more passionate about their work if they are treated with respect and given an opportunity to blossom – if the organizational leaders let them know "they matter."

One Leader's Experience: Are You a Spark of Light?

In a blog post, Tom Hopkins, Continuous Improvement Specialist at a very large distribution company, captures what a positive spark looks like.[3]

> I was helping one of our sites during our Peak season. Peak is the time of year where our people and processes work at their most stressed level of performance. I've been thinking a lot about the idea of a spark, and I've witnessed what it's like when a spark catches fire within a team …
>
> I met a manager (a spark for sure) who created a stable environment is one of the most hectic areas of a shipping plant – the dock (or shipping platform). His leadership style is disciplined and accountable to a process. He practices the skills needed to manage at Gemba, whether he was formally trained in this or not. He strives to live by the principles of humility and respect – and openly admits that sometimes he may let emotion get the best of him. This shows me that he has some ability to self-reflect. What I found most interesting about him is his mindset of support for the team, the process, and making things easier for everyone.

On my Gemba Walks with Eric, the dock manager, we saw the process, we saw the variation in the process flows, and we saw how he tried to bring it back to a standard – that is, back from gridlock and into a flow again. During our walks, I watched how he interacted with his supervisors. Instead of just barking an order (which he did at times), he would typically provide the thinking behind it. When there was more time to make a move, he would ask his supervisors questions, about what they saw and thought should be done. These questions helped him see what they saw and gave him the opportunity to coach them. He supported their development, while also holding them accountable to a process. He would walk a certain route at different times of the night/morning and would easily interact with nearly all his employees during these walk routes. Each time he would pass his supervisors, he would check up on the process, try to find where they were struggling, and give them support.

What I noticed was how those interactions changed the attitudes of his supervisors. There was so much interaction and learning going on each day; you could easily see a difference in them compared to supervisors who didn't get that type of interaction. Seeing how Eric's supervisors smiled, talked with people, and made decisions was amazing. Not all their decisions were correct, but Eric didn't reprimand them for it. He explained (rather sternly at times, for sure – remember, he knows he can be this way!) what those decisions meant to the overall process flows and to the customer. Just seeing his supervisors making those decisions differed completely from some of the interactions we would see elsewhere inside the building, where many supervisors seemed afraid to make a call on almost anything.

Build Relationships – Things to Think About

A Personal Story

Permit me to take those thoughts and share one of my key life's learning experiences.

In the mid-1990s, I was just getting involved with Toyota Motor Company type improvement practices. Prior to this point, I'd been actively involved with Total Quality Management (TQM) and what I thought were relatively holistic process improvement activities. Then I experienced something that radically changed the way I approached improvement. I saw something that I'd never seen before!

I had run a Kaizen workshop for a manufacturing company in Shreveport, Louisiana. A Kaizen team (U.S. style) is typically a one-week improvement project. "Kaizen" is a Japanese term meaning "change for the better." The team had done an outstanding job! We had completed our presentation to management, successfully implemented several changes to a complex manufacturing line and to their changeover process. Everyone was very pleased with the work done by the team.

When we returned to our meeting room to wrap up our activities, I asked the team members a question: "You spent an entire week working on this project. What did you get out of this investment of time?" We went around the room, and everyone said something polite. Then we came to Pearlie, and she said something that changed my life.

Pearlie said, "I've worked for this company for twenty-five years," she paused, then continued, "It's the first time they ever asked me to think. And I really liked it!"

When Pearlie made her statement, it took me to a new level of understanding employee engagement and how it was badly lacking. Previously, I had a good intellectual understanding of why we should engage people. But I did not have a deep emotional understanding or see the pain this lack of involvement caused. How could Pearlie be highly engaged at work if no one *cared* about her thoughts? If no one was interested in what she knew. If no one *cared* to help her develop her critical thinking skills.

It was the first time I emotionally realized how much leaders were missing by not doing this – *I learned to "see it!"* Pearlie gave me a gift, and 30+ years later, I'm still very grateful for it.

That experience changed the way I worked with future clients regarding engaging people, both on my project teams and in coaching leaders. Since that experience, I've continued to improve my ability to see things that were previously invisible to my eyes. My primary goal in writing this book is to help you have a similar experience.

Employee Engagement

At the very beginning of this book, we noted surveys from Gallup, Wyatt, and others commonly show less than one-third of employees are passionately engaged in the work they do. *And the number one cause of disengagement?* Surprisingly, it's not senior leadership. If it's not the senior leader, then who is it? *It's the employee's immediate boss!*[4]

Jim Clifton, CEO at Gallup, in his 6/13/17 Chairman's BLOG wrote[5]

> While the world's workplace is going through extraordinary change, the practice of management has been frozen in time for more than 30 years." According to Gallup's World Poll, many people in the world dislike their job and especially their boss. The Gallup Poll noted, "the current practice of management (in Japan) is now destroying their culture – a staggering 94% of Japanese workers are not engaged at work.

Employees everywhere don't necessarily hate the company or organization they work for as much as they dislike their boss. Only 15% of the world's 1 billion full-time workers are engaged at work. It is significantly better in the United States, at around 30% engaged, but this still means that roughly 70% of American workers aren't engaged.

What does this mean for reversing the slowing down of world productivity trends? *It means we need to transform our workplace cultures. We need to start over!* To summarize Gallup's analytics from 160 countries on the global workplace, our conclusion is that organizations should change from having command-and-control managers to high-performance coaches.

What the whole world wants is a good job where they can do meaningful work, and we are failing to deliver it – particularly to millennials.

Looking at LinkedIn, Reddit and other social media you will sometimes see a person asking, "I have an improvement idea, should I suggest it to my boss?"

The vast majority of responses are typically negative. People say:

■ "Complain to no one ..."
■ "Only compliment the good things ..."
■ "Try not to criticize management and you will be considered a team player."
■ "If you suggest an improvement, your boss will be threatened by you?"

Very few comments were made suggesting this person share their ideas. No wonder people lack engagement and enthusiasm about the work they do. How depressing! But it does not have to be that way.

While we might all wish we had an enlightened CEO driving improvement, the current reality suggests that most CEOs are not actively engaged in improvement activities. They want it to happen, but their attention (for better or worse) is focused elsewhere. So where does that leave us if we have a disengaged team or a disengaged department? From my perspective, *it's our responsibility as leaders to do something about it.* That may require some behavior changes since it's much easier to act as a "manager" than it is to act as a "leader." Think about those words for just a moment. What are the definitions for these terms?

- **Manager:** A person responsible for controlling or administering all or part of a company or similar organization.
- **Leader:** A person who guides and seeks to maximize the efforts of others toward the achievement of a greater good.

They certainly have some elements in common, but there are also significant differences. "Manager" implies controlling what is being done right now, with much more emphasis on the manager making sure work is being done in the right way, right at this moment. "Leader" has a more forward and holistic-looking implication. Both want the "right thing" to be done at the "right time," but the perspectives differ significantly.

Do your peers, subordinates, and your bosses look at you as a leader or as a manager? It would seem to me that "leader" has more positive connotations. So perhaps we need to change the way we are thinking, maybe even ban the use of the word "manager." You don't have to be handed an official leadership title to influence others around you. You can be a leader in any role.

Tom Hartman (retired senior executive from Autoliv) said, "Most people when they first become a leader are achievement-oriented. But the most successful leaders over time transform from 'what *they* personally can do to impact the organization' toward 'impacting the organization through others' (i.e. developing people, mentoring, etc.)." Tom realized that the only way to make an impact that is sustainable is through the continued efforts of others. It was a consistent insight shared by many others, including the story Jess Orr shared in the Reflections chapter (Chapter 2).

Most people begin their day in neutral and then react to whatever confronts them. That is one way to get through a day. If you are a leader, that approach is insufficient; you need to do more than muddle through. Your primary responsibility should be growing people, and helping them to recognize and use their talents.

Bob Chapman, retired CEO at Barry-Wehmiller, likes to tell the story about one of his employees who said, "When I go home at night, now I can talk to my wife." We don't want to be sending people home at the end of the workday who feel small and undervalued. Every job is important; if it were unnecessary, we shouldn't be doing the work. We need to help people understand the purpose of the work they do and then think of ways to engage them in improving that work activity. Allow them to give more meaning to the work they do and the way they do it.

To make a difference in someone's life, you just need to care. The transformation of people is amazing! Even if they leave your team, you will still benefit; if you do this regularly, you will be a magnet to attract new talent.

Find That Hidden Treasure

It begins by changing *"what"* leaders believe, and then changing *"how"* they lead. Do you really care about the people with whom you work? Do you believe your primary job is to get your work done through others? Do you believe your job is to create new leaders? I certainly do.

Steve Jobs once said, "Management is about persuading people to do things they do not want to do, while leadership is about inspiring people to do things they never thought they could."[6] Seventy percent of the work we do, people see it as part of the job, and they are willing to do it. But 30% is optional, *and that is where the treasure lies.* That treasure is available for any leader at any level of an organization to discover – if they have the discipline and tenacity to open their eyes and find it (Figure 4.1). We need to move from managing to leading.

These belief changes partly come from reflection and changing your perspective. And as you redefine your purpose, you are then in a better position to begin to change your behaviors. This will not happen overnight. You need to practice and get feedback along the way. The four foundations reinforce one another. As you practice executing the various actions described in this book, you build better relationships; you both listen to and coach the people on your team. You gain insights, build new skills, and you learn as a leader.

Leaders in companies who are highly effective at improving all seem to have one common trait in how they behave. They lead with a high degree of humility and a willingness to learn. As these leaders grew from leading teams, departments, and organizations that were good at improving to a much higher level of improvement maturity, they radically changed their perspective regarding how to lead.

They created an environment to tap into that treasure where people learn, where they improve their critical thinking skills, and where people

Figure 4.1   Hidden treasure

feel they can influence the way work gets done – for the better. These leaders want every "Pearlie" on their team to blossom! Almost all leaders realize they need to nurture their high performers. Outstanding leaders also realize how important it is to take care of their dependable performers and help them to grow. Pearlie absolutely had passion, she had lots of it, but the company and her direct boss made very little use of her talents and capabilities. The most effective leaders elevate average performers.

Learning as a New Leader

When I had my first leadership position at Beatrice Foods, I was placed in charge of a new team. I was able to hire all of the team members. Two things happened during my first couple of years as a team leader that caused me to alter the way I led and made decisions. The first lesson was somewhat subtle. The second lesson took me quite a while to realize.

I hired Ralph; he had worked for a Big 8 Public Accounting firm for five years. He was a very intelligent guy and turned out to be a really great team member. Ralph had been on our team for more than a year when he came into my office to have a conversation. He said, "I don't know what you do, but whenever I have a problem getting something done *(working with other groups in the company),* after I talk to you, the problem just goes away."

At first, I thought this was pretty cool and that I was doing a good job as a leader. But then, over the next couple of days, I reflected on what he had said. I realized he was really saying, "Why are you solving all of my problems? How will I ever learn?" It had never occurred to me that I might be doing the "wrong" thing. I had pretty good relationships around the company, and it just seemed natural for me to get the obstacles removed for my team members so that we could get the job done more quickly. *I did not see how I was inhibiting the abilities of people on my team to grow their own capabilities and figure out how to address those challenges on their own.*

That conversation with Ralph changed the way I led. I would still intervene if one of my teammates asked me to do so. Rather than going to get the problem fixed myself, I then started practicing asking better questions; this was my early attempt at becoming a coach. Instead of clearing the obstacles, I began to converse more with my team members on why the assignment they were given was important and how it might fit with other activities underway at Beatrice. When they became more informed about the "why" we were doing something, they became much more comfortable having those conversations with other departments, and there was little need for me to run interference. The other functional departments became much more cooperative and collaborative when they did their work. It took me a while to get good at asking questions; I'm too embarrassed to say exactly how long ... but let's just say it took more than one year of practice to routinely start asking *good* questions.

I was not initially smart enough to fully probe what Ralph was "really" saying when we had this conversation; it required some further reflection. The second learning stayed hidden from me for quite a few years. I always pictured myself as a very inclusive leader. I was always open to improvement ideas and always found "change" an exciting thing to do. But what I did not realize was my need for "control." In all my leadership positions, I deeply wanted to make sure we accomplished whatever goals were given to us. I didn't need to do it, so the light shined on me; I was good at giving credit and having many of my teammates promoted. But if our job was to take the next hill, we were going to take the next hill.

I learned the second part of Ralph's message five years later when I became the Director of the Information Systems Group. When I took over that responsibility, our group could guarantee with 100% certainty that we would not give you what you asked for, and we would give it to you late! One might have said, "Our track record was consistent and totally unacceptable."

My challenge and what needed to be fixed were quite obvious. My only problem was that I had been out of the information system's world for seven years. Technically, I was a dinosaur. I had no feel for the technical decisions we needed to make, and I was terrified! I realized I couldn't possibly control what was happening, even though we had serious issues. My learning opportunity for this challenge was to learn how to completely trust people!

Until that point in my career, I did not see my "trust and control" problem. I had been given this position because clearly the department needed to be fixed. For the last four years, I had been responsible for leading a company-wide productivity improvement effort. We had done a reasonable job of improving within this diversified global conglomerate of more than 440 different business units. Beatrice was one of the sites visited when the Malcolm Baldrige quality award criteria were being created. I'd been put in charge of the department because, in theory I knew something about improvement.

I now got to practice what I'd been preaching for four years. That was also a little scary. My first several weeks on the job were spent talking with people on the staff and with the customers we served, all of whom were internal departments. There was a long list of concerns. Our biggest problem was that no one trusted us; most of the internal departments wanted to work with outside vendors. The technology was also changing, and personal computers were beginning to be extensively used throughout the enterprise.

The previous leader of the group was a good and knowledgeable person, but he operated like the IT Police Department and was very much into controlling what was going on. Given the advent of personal computers, those days of "rigid" control were over, so it was clear that change was going to happen with or without us. The way we developed new information systems was also an obvious part of the problem, resulting in late deliveries and delivering something that was not desired. The picture shown in Figure 4.2 captures our old process.

The customer requests some type of report, a program, or the creation of an algorithm. The work is passed along from functional group to functional group. At each step of the process, people go back to the customer to get a

Processes often look like this...

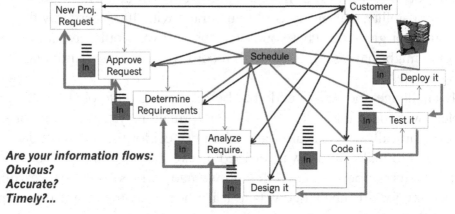

Are your information flows:
Obvious?
Accurate?
Timely?...

Figure 4.2   Erratic information flow

redefinition of the requirements. In the 1980s and 1990s, it took so long to get a large program developed that by the time the request was fulfilled, it was highly likely the original customer for the project was no longer there. So, you might imagine the chaos inside such a development process.

We changed that model. We developed four rules:

1. No project can take longer to develop than one year (in today's world, I would seek much, much shorter cycles).
2. We would do a mock-up of the request on a personal computer to validate the requirements.
3. For routine programming tasks (e.g., print protocols, screen management, etc.) we would develop a shared library of reusable code for use by everyone doing software development.
4. Projects that were subject to audits by regulatory authorities had more restrictions and requirements development protocols; outside of those types of projects we would try to do whatever the customer wanted in the most simple and reliable way possible.

Our internal customer groups approved the rules. We also had an additional internal goal, "to deliver bug free software, by the requested date." We did not initially share this with our customers because we had no credibility. I was working with the same staff that had been in place for the last several years. Within two years, the team turned around our relationship with our customer groups, and we regularly met our delivery and quality goals. Customers began trusting us more and would bring us into the conversation earlier when they were trying to decide what they wanted to do. It was great!

I learned something significant about myself in this role. I did not realize how much I relied on thinking; "I could control a project as well as or better than any of the other people on my team." I'm not saying I led that way, but I very much enjoyed believing it. In this situation, I did not have the skills or technical ability to control the projects we were doing.

What I learned to manage were the four rules. Project leaders often requested my participation in key customer meetings. Observing how the teams interacted with our customer groups was insightful and helped me gain insights on how to coach them in their project management activities.

Interestingly, my letting go of control also required learning by my staff. We had one breakthrough moment when the leader of Systems Development and the leader of our Database Management group came into my office to ask for a decision on something they were energetically discussing. After listening to both people, I said, "I do not know what you are talking about, but if you want me to decide, I'll do it." Their eyes got big, and Bob said, "That's Ok, we'll take care of it."

That discussion quickly spread amongst the staff. The outcome was that people in the department started taking more responsibility for working together. Your teammates cannot score a point if you refuse to give them the ball. *Letting go of control as a leader is a requirement for growing people.*

A question for you as a leader might be, "What rules are you following in the way you lead, and are those rules clearly understood and agreed upon by your teammates?"

Letting Go of Control at Autoliv to Get More Engagement

One of my more romantic notions of how improvement happens was tossed into the dust during my research. Autoliv's leaders worked with a highly respected Toyota Sensei early in their journey – Harada San. I've always assumed working with a person like Harada San would somehow immediately turn on the light, the road ahead would be obvious, and transformation would "magically" take place. Intellectually, I knew this was unlikely, but I always wanted it to work that way. Alas – it does not!

Getting it done is mostly hard work, a little bit of serendipity, and the tenacity to keep plowing ahead even when the road seems quite murky and progress seems slow. It turns out that the importance of having the humility to let go of control is also key.

Autoliv manufactures airbags for the automotive and trucking industries. It is a regulated company working in a very competitive industry. Auto manufacturers typically expect 5% cost reductions on an annual basis. When we talk about "letting go of control" it causes a tightening in the chest for many leaders. They wonder about quality, on-time delivery, and cost management if there are "no" controls.

Let's clarify what this expression means before proceeding further. Letting go of control does not mean chaos and a lack of accountability. What it means is processes have been developed that are sufficiently reliable to produce quality products, on time, and in both a safe and cost-effective way. It means the leader is willing to be in touch with the actual reality of what is happening within her area of responsibility and with her customers. And it means the leader has a large dose of humility where she can listen to other viewpoints, so that both she and her associates are learning.

Early in their improvement journey, Autoliv was very focused on getting improvement ideas from their associates. That was initially how they measured engagement. A Kaizen at Autoliv is simply an implemented improvement idea. The specifics of the graphic below are not important; I want to focus on two significant changes by the leadership team relative to "control."

12,000 implemented ideas ... good, right?
The bars on the far left in Figure 4.3 were ideas implemented by associates in 2005; it was approximately 12,000 ideas from a workforce of roughly 3,000 people. A good number for most companies, but not satisfactory for Autoliv's leadership. The *leadership team realized they were a bottleneck relative to implementing "good" improvement ideas.* Many of the improvements employees wanted to make did not impact the product or process integrity (e.g. move this fixture from here to this spot closer to my hands). Impressively, leadership decided to stop being responsible for approving ideas that did not impact quality or process integrity.

Could they do more than 12,000?
The way they did this was both simple and elegant. After 2005, if an associate wanted to implement a change, the first thing she needed to do was to get the people on her team (her shift) to approve the idea. If they did, the supervisor was responsible for communicating the idea to the next shift during the daily transition meeting. If the other shift approved the idea, it was

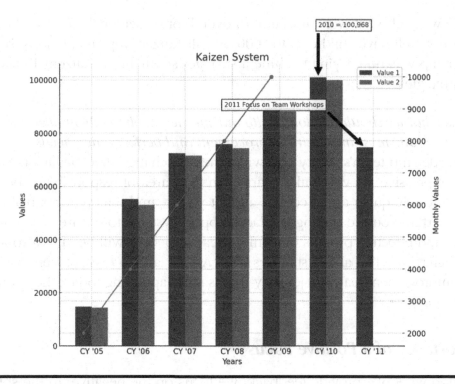

Figure 4.3   Kaizen ideas implemented

implemented with one exception. Manufacturing engineers were assigned to support the cells. When they reviewed an improvement idea, if they felt it might impact quality or process integrity, they would first communicate with the team to get a better understanding and, if they still felt a higher-level review was necessary, those ideas would get submitted for management approval.

Following this new practice, Autoliv increased implemented improvement ideas (Kaizens) from 12,000 to more than 100,000 in 2010. In 2011, leadership decided another change was needed.

100,000 as good as it gets ... right?
Although the leadership team was feeling pretty good about employee engagement, they wondered if they could do better. They decided to change the metric from implemented Kaizens. They still wanted associates to implement their ideas, but the only Kaizen implementations that were going to be measured moving forward were Kaizens that touched one of the organization's Policy Deployment (key strategic) goals. The number of implemented Kaizen dropped by about 50% to 50,000. This shift helped to make everyone

more aware of what was important to overall organizational success. There may very well have still been 100,000 overall Kaizen implementations, but leadership was now trying to shine a stronger spotlight on Kaizens linked to key strategies.

This is what we should be doing with our metrics, evolving them over time to align with our most important improvement and performance needs.
The leadership team's ability and willingness to change their behaviors were quite impressive. As you make changes in your area of responsibility, be willing to relinquish direct control and let your team be more accountable. Make certain you are putting in place support systems to ensure account-ability and effective process outcomes. Realize people will need to also learn new behaviors. The above statistics from Autoliv are an excerpt from a 15+ year improvement learning journey that is still happening today; they continue to evolve.

Importance of a Positive Focus

So often when we provide feedback, we focus on the negative. In the spirit of changing your perspective, think about doing the opposite. Tom Landry turned around the Dallas Cowboys in the 1970s and 1980s. One of the stories attributed to how he accomplished this dealt with how he provided feedback. Other coaches focused on missed tackles and dropped passes. Landry realized there are many ways to get something wrong and only a handful of ways to do something right. So he studied film from prior games and gave each player a highlight film of when they had done something right, something effective. Landry felt the best way for a player to learn was by looking at their performance in slow motion to see what they had done right; their own personal personification of excellence.

When you see someone doing something right on your team or in your organization, share what you observed. Let them know you saw excellence and tell them how that made you feel. Don't say, "Great job!" Instead, be specific with what you observed. You might say, "I was really impressed by the way you XXX." This type of feedback isn't a judgment, it exhibits humil-ity by the observer and provides powerful and positive reinforcement to the person doing the work.

Kevin Meyer posted a note in his blog[7] on the power of positively focus-ing on our successes, starting with a quote from Taiichi Ohno:

Most of us look for reasons when we fail, but very few of us look for reasons when we succeed. *It is important to search for the reasons why you were able to succeed, and make sure to use the acquired learning in the future.* Not only should we reflect on success in order to continue to improve on that success, but we should also deliberately reflect on success so we can learn and apply those lessons to other activities. [emphasis added]

The deliberate in *"deliberate reflection"* is the key. Deliberate (or "intentional") reflection is not simply thinking about a circumstance or situation, but actively planning for and methodically executing the reflection process itself. In fact, some of us who have come to embrace deliberate reflection will deliberately reflect on the reflection process itself. Like any process, it can always be improved. I try to practice deliberate reflection daily, weekly, monthly, and annually – all with different formats and purposes.

Preparing for deliberate reflection includes deciding on a location with appropriate surroundings and stimuli (or lack thereof), the time required, focus or topic, questions to be asked, and desired outcome. Questions always include some form of "what is the desired future state," "what is the current state," "what are the gaps," and "what should I do differently?" And of course, "what am I doing positively that can be a foundation for future success?"

Mike Robbins quotes a story from his conversations[8] with Erica Fox, who was at the time the Head of Learning Programs at Google, on the power of a positive focus. In a LinkedIn blog Mike said,

> [S]he came up with an idea of how to engage her direct reports in a positive way. Since she was leading a remote team of people who were in various cities, it was challenging for them to connect in a personal way. Even with the use of Google's state-of-the-art video-conference technology, there's nothing quite like being in the same room. And as anyone who leads or is part of a team that is distributed across multiple locations knows, it can be difficult to connect effectively and personally via conference call or video conference.
>
> During a weekly meeting, Erica asked each of her team members to share something they were grateful for from the previous week – it could be something work-related or something personal, so long as it was something that they genuinely felt grateful about. She asked them not only to share this verbally with their teammates, but also

to write down what they were grateful for on a Post-it note and stick it somewhere out of sight in their work – space (like inside a folder or desk drawer). She thought it would be fun for them to find the Post-it note again sometime later and be reminded of the positive thing they were grateful for that they shared with the team.

The exercise was fun and set a nice tone for their weekly team meeting that day. It allowed people to connect with one another in a more personal and positive way, even though they weren't all sitting in the same room together. It went so well the first time she tried it, she decided to do it again the following week. Some of the people on her team were more into it than others, which is often the case for things like this. She did it a third time in their next weekly meeting. She decided not to do it the following week because she thought it might be getting a little old, and she wasn't sure if the people on her team were all that into it. But when she started that next meeting without doing the gratitude exercise, to her surprise several of her people got upset. They had been ready with their Post-it notes and had already planned what they were going to share. So, she decided to do the exercise again that week and made it a standard practice for her subsequent weekly team meetings, which helped improve their personal connection and team culture even though they didn't all work together in the same location.[9]

Medical research has shown the power of positive feedback and I referenced this idea in the chapter on Reflection (Chapter 2), where the Dr. Dean Ornish story was shared about "Change or Die." People will not consistently change their behaviors if they believe the overall outcome will not be positive. Dr. Ornish's team provided support systems to help his heart surgery patients change their behaviors.

Another related piece of research on this subject was done by Dr. Elias Porter and his wife, Dr. Sara Maloney Porter.[8] They published an instrument based on their research titled, "Strengths Deployment Inventory" that is still available for people to use today. The thing I found fascinating was how they defined a "weakness." A weakness is often a strength that was used excessively. Thus, action-oriented and decision-making can be perceived as a strength, but if carried to excess it becomes domineering and poor listening skills. Great analytical skills, if used to excess, become analysis paralysis, and supportive/nurturing behaviors, if used to excess, allow others to "kick the dog" as a group of air force officers said in a workshop many years ago.

So positive reinforcement is powerful and is underutilized, but you don't want to do it to the point of being a Pollyanna and out of touch with what is happening. One of my favorite movie lines was written and spoken by Orson Welles as Harry Lime in *The Third Man* (1949).

> In Italy for 30 years under the Borgias they had warfare, terror, murder, bloodshed – and produced Michelangelo, da Vinci, and the Renaissance. In Switzerland they had brotherly love, 500 years of democracy and peace, and what did they produce – the cuckoo clock.[10]

Healthy debate and discussion of different viewpoints are valuable, but it must be done without demeaning or attacking the other person.

Power from Diversity of Thought

This is the subject of frequent discussion, and once again, intellectually, I always believed this to be true, but I'm not sure if I deeply understood it. I recently gave a speech on Artificial Intelligence (AI), which is a technology making significant changes to the way we do work.

One of the insights gained while preparing for the AI talk was a deeper understanding of the power of diversity. AI companies are developing powerful algorithms for analysis and, ultimately, for computers to make decisions. Companies in this field have learned if they do not have a diverse team of people (races, ages, ethnic backgrounds, experience, etc.) doing the coding to develop the algorithm, then they miss very important perspectives. One of Google's first experimentations with facial recognition had problems recognizing people with dark skin. This was largely due to the fact that the images initially loaded into the database were of people with white skin. The team did not do this due to blatant discriminatory practices; they simply did not give thought to the issue. This is often referred to as unconscious bias. A similar thing happened with a software program learning about jobs and when doctors were referenced, primarily white males had been loaded into the software, so if a female or minority picture was looked at, it was more likely to be referenced as a nurse or a technician.

Where diversity is encouraged and accepted, it builds trust between individual team members. This is especially important when we have teams of people spread across the globe and innovation is critical. Diversity of thought is a great way to change one's perspective on life.

People carry an unconscious bias based on their experiences and the values they develop. If leaders can help to change people's perspectives and help them overcome their unconscious biases, then their teams can do more effective work. The mere fact that you, as a leader, create an environment where diversity of opinion is encouraged and debated in a healthy way will go a long way toward improving the effectiveness of your team. Regardless of whether you are leading a cell in a manufacturing plant, a project team, or an R&D department.

Two of my favorite companies in this area are W.L. Gore & Assoc. (specialty materials manufacturing) and O.C. Tanner in Salt Lake City, Utah (works with clients to increase employee engagement and improve business results).

W.L. Gore & Assoc.

You are only a leader at Gore if people decide they want to follow you. Every day you need to prove yourself. Leaders are regularly assessed by their peers and followers based on five dimensions of leadership that include: Leading self, leading others, shaping the vision, getting it done, and living the culture. They have used a "Trust Index" tool for senior leaders that includes the following questions:

1. I feel he/she creates an environment of trust.
2. Culture is valued as a critical means of achieving business results.
3. Accountability and decision-making are clear within the division.
4. Associates and teams feel empowered to make decisions.
5. Associates can speak their minds without fear of retribution.
6. I feel he/she seeks appropriate input when making decisions.
7. Diversity of thought and perspective is encouraged.
8. I feel he/she creates a healthy balance of challenge and support.
9. I feel he/she is approachable.

Leaders do not decree actions within the Gore world. They need to explain their reasoning, and the topic gets debated. The organization believes that while this may initially take more time, in the end, the actions are better, and there is more ownership of them because of engaged associates. Implementation is faster and better because of engaged, passionate people in the planning stages.

O.C. Tanner

Salt Lake City is a refugee resettlement city; they encourage refugees from around the world to come to their city. Associates who work at O.C. Tanner are a melting pot of people; they speak over 40 different native languages. Whenever they hire a new person, on that individual's first day of employment, they are encouraged to bring something to the workplace that celebrates their culture. This gives their teammates an opportunity to recognize the background of their new team members and to appreciate the diversity that individual brings to the team. O.C. Tanner is a wonderful example of how people with different religions, different ethnic backgrounds, and different nationalities can come together and work as a high-performing team. Leaders and associates all show an obvious respect for one another.

My Values, Your Values, and Reality

Values are ingrained into us from an early age. But just because we believe something is true does not make it true. Partly, we can test the validity of our values and beliefs by reflection and looking for evidence that reinforces the belief or evidence that clearly states that perhaps there is a different reality. In the United States, many people claim to believe the earth was created 10,000 years ago. They turn a blind eye to the fossil records and fail to grasp that if they walk outside and stare into the nighttime sky, they can look at the star Cas in the constellation of Cassiopeia. That light took 16,308 years to reach our naked eye. If they look at the distant galaxy Andromeda, the light took 2.5 million years to reach the Earth. *"Facts" do not cease to exist because a person chooses to believe something else; the denial of basic scientific facts limits a person's critical thinking ability and, for leadership purposes, makes it difficult to accept values that may be held or believed by other people from other cultures.*

There are certainly wrong things we can do in life. Encouraging anyone to harm another person and telling lies that injure the reputation of another person are totally unacceptable and illegal behaviors. When we discriminate against another person, we lose the power of diversified thinking, and in the global world of today's economy, we lose the ability to better understand a diversified set of customers. In that type of toxic environment, even people who are "like us" become more reticent to speak out for fear of being ostracized. So, *truth gets hidden.*

There is power in interacting with people who speak different languages and who come from different cultures because it will help us to see more possibilities. In the short term, there may be language or cultural challenges as we try to figure out how to work effectively with one another. But as time goes by and trust is built, our fears of the unknown shrink, and we realize people from other cultures are just as human as we are. That is one of the things so wonderful about the O.C. Tanner culture when they celebrate and recognize the differences on the first day a new associate starts to work there.

This openness is contagious in the most positive sense of the word. When the leader and organization are open to the viewpoints of other cultures, they are also open to my viewpoint and to your viewpoint. Operating this way as a daily norm moves us toward that spirit of openness where people are comfortable asking difficult questions, and the unit/team/department is interested in understanding their true current reality.

Bill Gates and Melinda French Gates were interviewed on the subject of values. They point out how, when they fund projects in other countries, they must take into account the values, viewpoints, and expertise of the local culture where they are doing the work. In their Annual Letter [11] (in answer to the question), "Are you imposing your values on other cultures?" they said:

> **Bill:** On one level, I think the answer is obviously no. The idea that children shouldn't die of malaria or be malnourished is not just our value. It's a human value. Parents in every culture want their children to survive and thrive. Sometimes, though, the person asking this question is raising a deeper issue. It's not so much a question about what we do, but how we do it? Do we really understand people's needs? Are we working with people on the ground?

> **Melinda:** We're acutely aware that some development programs in the past were led by people who assumed they knew better than the people they were trying to help. Over the years, we've learned that listening and understanding people's needs from their perspective is not only more respectful – it's also more effective …

> We have about 1,500 employees. One of the most important parts of their job is listening to partners, adjusting the strategies based on what they hear, and giving implementers the leeway to use their expertise and local knowledge. That's not to say we always get it right – we don't. But we try to approach our work with humility about what we don't know and the determination to learn from our mistakes.

Leaders Lose Credibility When Processes Are Not Stable

On the Netflix website, they talk about the danger of becoming a "process"-driven company. Implying that process-focused organizations restrict creativity, inhibit learning, and are overly constraining.[12] I agree if they define the word "process" as inflexible rules.

But think about this for a moment. You are trying to get your work done and the information you need is regularly not available, or the parts are not there to do the work, or you need a decision to be made but it's not clear who has the authority to give the green light. What do people do during these situations? Typically, one of two things happens. You wait, or you do a workaround and try to get the job done using some alternative way to complete the task.

Every time people on your team do a workaround, you lose credibility as an effective leader. Often these workarounds go unnoticed by the leader because workarounds have become the standard way work gets done. This is an unacceptable situation if you wish to be highly effective. The leader also sends a message to his team: "I know we have these problems, and I am powerless to do anything about them".

Figure 4.4   Workaround

Your workarounds look like this (Figure 4.4).

- But usually, we don't see pictures of them.
- They remain hidden because the work gets done.

Unfortunately, most organizations do not have a standard work protocol to focus a spotlight on workarounds. People are typically too busy getting the work done. The processes they use are rife with variation, and no one has time to fix them. As time goes by, we (as leaders) cannot see the work-arounds happening because it becomes such a normal part of the day-to-day environment. We accept it, but we shouldn't!

This is also one of the hidden conundrums that makes it difficult for organizations to sustain the gains from improvement activities. Every time a workaround gets done, it introduces variation to the process. Processes with significant variation are unstable from a statistical perspective. They increase the likelihood of quality problems with the work being done. It does not matter if it's a part being manufactured, software being coded, or physicians using different procedures for the same operation.

What should you do if unstable processes surround you? You need to decide! Are you going to live with them? Or can you find a tactful way to raise awareness of what is happening? At the outset, you probably need to do both. Depending on your situation and priorities, you may need to ignore some issues and spotlight others.

One thing that can help is using "Visual Leadership" to make problems more visible. We will discuss this in Chapter 5; visuals are powerful way to inspire action.

Another action you might try is to do something similar to a story shared by Steven Spear in "High-Velocity Edge"[13] and in a presentation I heard at an AME Annual Conference. I am paraphrasing the story as something similar has happened in several healthcare institutions.

Steven shares a story about a patient who went to the hospital for a fairly routine procedure, but she died during her stay because of a medical error. A nurse meant to administer one drug to flush a "central line." Instead, and accidentally, the patient received several doses of insulin that cratered her blood sugar and left her in an unrecoverable state. How could something like this happen in an organization with so many smart people dedicated to a noble purpose?

The hospital had unreliable standards for doing the work. The standards lacked the clarity to make it easier to succeed and harder to fail. When people had difficulty doing their work, they were not encouraged to point it out or to inform anyone about the issue. Nor was anyone regularly observing how work was done and the conditions in which the nurses performed their work.

This incident couldn't be the first time a nurse picked up the wrong medication vial. Previously, they'd caught themselves and then grabbed the right vial. But was this looked at as a failure of their standard? Was there ever a conversation about why the wrong vial got picked up? Did the nursing staff feel their improvement thoughts were welcome?

When the above issue took place, the mistaken vial was identical in color to another medication. The labeling on the bottles was in small print and quite similar to the wrong vial. The bottles were close together in the medication cart. The incident happened at night when the patient was sleeping. The nurse was working in a room with a very low level of lighting so as not to not disturb the patient. There was a laundry list of minor problems on the night this happened, and they resulted in a terrible tragedy. The standards within the organization for pointing out and acting upon problems were unreliable and arguably inadequate.

Unfortunately, this tragedy is indicative of an issue in many organizations – "Management's behavior and the way we approach improvement." There are many opportunities for leaders to observe how work gets done. Administering pain medication to patients is obviously an important process at a hospital. How might the tragedy have been avoided if the way work was being done was periodically observed by a leader looking at a work activity where the nurses (in this example) had previously pointed out difficulties in getting their work done?

It takes effort by leaders to build trust so their team members are comfortable pointing out process issues.

What is the learning *for us* from this terrible tragedy?

At the start, you probably have a lot of processes that are not stable, with variation happening daily. You can't work on them all at one time. Determine which processes are the most important priorities given your current situation. Considering the medical story above, the most important processes for a nursing staff on a hospital floor that a leader could observe might be:

1. Administration of medications to patients.
2. Timely and ease of charting and updating patient records.
3. Timely and appropriate rounds during the shift, making certain protocols for handling issues that arise with patients during their stay are addressed.

Assuming those were the three most important three processes, then every time a workaround is required for those processes, or if the process fails to operate properly, or if there is a near miss, there should be an action taken. Nurses must be comfortable and believe the issues they point out will be addressed. Ideally, several people can participate in observing the problem steps within the process. Over time, the process will become more stable, and the need for workarounds will decrease. Key processes like the above could also be part of the leadership team's Gemba Walks when they are observing the effectiveness of standard work practices.

When making these observations, you will not always find a root cause, but sometimes you do. Importantly, you are making variations in the process more visible by the mere fact that several people come to see what is happening. It makes everyone in the area more aware of the importance of stabilizing process performance and driving out unnecessary variation.

A process should be the best way we "currently" know how to do something. And if that is the best practice, how can we get more people to practice doing it? Realize that even what we call a best practice is still "an experiment." *It is the best way, at this very moment, that we believe to be true for doing this activity.* In the spirit of improvement, we should continually conduct new experiments to improve our best practices (processes) and take them to the next level.

I would argue that being a process-driven company does not restrict creativity. The discipline of trying to understand and execute work activities in the most effective way can create a sense of freedom. When you become a process-centric thinker, you are not doing it to make sure the rules are followed. You are doing this to better understand how customers are being served from standard work practices and activities that typically go across several different departments or work groups. Some of the ideas discussed in the Visual Leadership chapter (Chapter 5) can increase the *visibility* of cross-functional process performance.

Where Does Trust Come From

Teams that work best together are those in which the members trust each other. Everyone agrees to complete their tasks, to say what's on their mind, and to respond openly and fairly to comments and criticisms. Each person understands excellent results come from everyone being involved. Conversations between team members might be intense, even heated, but

Figure 4.5 Trust model

if people treat one another with respect, they can learn together and create a better outcome. A diverse set of perspectives is welcomed and used as a learning opportunity.

As Figure 4.5 shows, trust requires three basic attitudes or behaviors between people.

Making and Keeping Agreements

A commitment to making and keeping agreements is essential to all successful relationships, both personal and work-related. To achieve this, fuzzy agreements are unacceptable; agreements must be clear, realistic, and achievable. When an agreement cannot be met, the people involved are notified, and actions are taken to get back on track or to revise the agreement. Dependability is a key foundation for trust. This means that work and performance must satisfy each party's expectations of a quality service or product.

Credibility

Making and keeping agreements establishes credibility. Credibility includes a relational dimension; what a person says and does is consistent and aligned. Ideas and opinions expressed by a person with credibility are deemed valid when giving feedback to another person. A person who possesses credibility is worth listening to because they offer ideas that move everyone forward.

Openness

People show each other a willingness to share information. Each person is receptive to others' ideas, willing to say, "I don't know," and interested in finding the "best" way of doing things, regardless of personal ownership. People who are open with each other are not afraid to get involved and see participation as necessary for success. Teams that can operate with a high degree of openness typically operate quite effectively.

Trust

When people relate to each other with openness and credibility and learn that each one can be depended on to make and keep agreements, there is trust within the relationship. Honesty, integrity, and respect characterize the feelings each person has for the others. There are no lies and no exaggerations – there's no need.

- Leaders learning to trust and work with others is a primary point of this chapter. This shift in focus for effective leaders partly comes from their personal learning experience. But we need to go further; there is also a shift we need to make in the way we do work. When you first start your career, it is important to prove your capability and to develop technical skills appropriate to your particular discipline. But you don't want to get trapped behaving that way once you begin to lead others. You need to make a shift.
- When Novak Djokovic was being interviewed during the 2018 U.S. Open Tennis Match, he said, "Every single day people have moments of vulnerability. In the face of adversity is when you grow. You need to believe in yourself."[14] As you gain skills and confidence you can help the people around you to do the same if you are willing to be vulnerable and give them a chance to grow.

Actions to Practice – Build Relationships

Changing Your Behavior as a Leader

When a problem arises, how quick are you to offer a solution or to be the person figuring out what went wrong? You are not going to be building deep relationships if you continue to behave that way. So how might you become a more people-centric leader by altering your current behavioral pattern?

Consider practicing these three behaviors:

1. **Pause and Reflect**: Give some thought to "purpose" when you have conversations, when you use an improvement tool, or when you launch a team. Of course, you want to solve the problem or effectively use the improvement tool, but what are you hoping will happen? Just pausing for a moment before you start talking to consider purpose can alter what you say. It fits with what we were talking about relative to finding ways to change your perspective and your behaviors as a leader in the previous chapters.

2. **Probe to Learn**: In my younger days I'd be so excited when I thought I knew the answer when having a conversation with someone else, I couldn't wait for them to stop talking so I could share my idea. Many people have written about how poor listening happens when people operate that way. If you want improvement to happen on a regular basis, you MUST find ways to slow down what is happening in your head. *One way to do this is to ask a question when someone says something. Probe to better understand.* In my younger days, my purpose was to demonstrate my capability and to be useful. I still want that to happen, but now I'm much more focused on helping other people grow. That is my primary purpose. I needed to align my behaviors accordingly. If you get good at doing this, the people around you will get better and better and faster and faster at solving problems.

3. **Ask Good Questions**: In the *How to Do a Gemba Walk* book, we stress the importance of walkers asking good questions. Sometimes, when we think we know the answer, we ask leading questions, or we ask questions that put people on the defensive. Classic Gemba Walk literature suggests you start with "Why" but an even better start is asking "what" and "how" type questions.

This is so important; I'll expand on the last idea above as the first technique to practice.

Learn How to Ask Better Questions

It is much easier to tell someone how to do something than to teach them to understand. But if you want to develop and engage the people on your team, you are responsible for helping them better understand and develop their critical thinking skills. Asking good questions is one way to do this.

Many questions people ask are "safe" and surface what we already know or what we want to believe. People may also ask leading questions (an answer you want to hear is embedded in the question) or close-ended questions (people answer with one word, typically yes or no):

- Do you like working on my team? (possibly close-ended or somewhat intimidating as the answer you want to hear is quite obvious)
- How hard did Bobby hit you in the face with his fist? (leading)
- Are you taking your medications as directed? (closed)
- Why do you keep having so many problems getting your work done? (leading)
- You do not miss any doses of taking your medication, do you? (leading)

The way questions do/do not get asked has a big impact on people. People at the worksite are also observing. Thoughts are constantly going through their heads: "What does the leader really want to know? If I share something that's wrong, what will happen? Does this person want to know what is really happening? Is it worth taking a risk and saying what I truly believe?"

One way to help create this environment is to focus on performance targets, with a process mindset that considers the suppliers and customers for the work being done. Organizations usually have overall future performance targets, but they are often missing a future state performance target at the department or work–team level. There should be two target conditions (goals) people strive to hit in the work they do. People usually focus on the first one and often miss the second:

1. What is the current performance target?
2. What is the "future state" performance target?

Appropriate questions to ask might be:

- Can you tell me why that is the target number?
- Does anything ever make it difficult for you to accomplish the target?
- Who are the customers for this process, and how do you know they are well-served?
- What has been learned from the actions taken?
- What are the next actions planned?
- Why these steps?

Great leaders ask good questions; they go deeper to lead us to inspiration and insight and demystify the unknown. They are open-ended or probing questions that encourage people to think and share information. As noted earlier, these questions often start with "what," "how," or "why." Questions AME Excellence Award assessors typically ask when doing a site visit include:

- What types of problems happen while doing this work? (open)
- How do you know when a problem exists? (open/probing)
- Can you show me an example of what you are describing? (probing)
- What do you mean by that? (probing)
- How do you know when things are running well? (open)
- Why do you think that happened? (probing)
- How could we learn more about what just happened? (open)
- How do you use this information to decide? (possibly open, borderline leading)

These are questions a leader might ask who is seeking to learn more. The leader is not trying to solve the problem; they are giving thought to "What is this person trying to accomplish here?" above and beyond the immediate task? There is a longer-term outlook to grow people and nurture critical-thinking skills.

Asking open-ended questions helps to increase understanding. Often, for both parties in the conversation. It is an investment in long-term change. It's helping others learn how to think more critically and to learn as they discover and gain experience. Asking good questions in a nonjudgmental way also builds trust.

As a leader, it's okay for you to have a repetitive set of questions you ask when talking with your team.

- So, tell me about what you are thinking?
- What is the problem you are trying to solve?
- How might we look at this issue in a different way?

Those types of questions help people develop their critical-thinking skills. They might even become part of your leader's standard work practices.

Questions for a Leadership Team

My good friend Mark Preston, in a LinkedIn post, suggested five questions for a leadership round-table discussion.[15]

1. **Any wins (big or small) this week?** This is a great place to start! Employees get to celebrate and even brag a little about all the positive things that happened that week by simply answering that question. As a bonus, you will understand what employees consider triumphs relative to the goals of the organization.

2. **What challenges are you facing?** *Without clear vision and under-standing,* we cannot receive the coaching and guidance that help us think about the issue in a fresh new way. Often, just speaking about our roadblocks helps clarify how to resolve them ourselves.

3. **How can "we" make "you" more successful?** Team member success is an always-evolving process. Sometimes your team needs more training or a one-on-one meeting. Other times they require help learning a specific skill set. This question gives permission to ask for the things that will move the needle forward and build a more engaged team.

4. **Provide one idea to improve the product or services that we are providing:** The experts are those who are closest to where the value is being created! The best source of innovation is often found by people who already work for you. One idea at a round table can be developed further with more minds. True – Kaizen, "Change for the Better" is best with a team. The team can transform the engagement and development of an average idea into an outstanding one!

5. **What were some great contributions made by other team members?** This opens the door for praise and team strength. By sharing what "great" looks like and what following company values looks like, the team will transform attitudes throughout the kingdom. By praising what "good" looks like, we are spreading a positive attitude.

Teamwork is not a given. You, as a leader, must continue to listen, question, and guide a spirit of positive inclusiveness.

Effective Team Huddle Meetings

Within the world of Lean or Toyota-type improvement practices, many teams hold a daily huddle meeting. These meetings are typically short, stand-up meetings that last from 10–15 minutes. They can be a complete waste of time and boring, or they can provide meaningful information and get people started on a positive vibe. Even if your organization is not conducting regular huddles, you might want to implement this with your team. To make your huddles effective, consider the following:

- Determine the purpose for hosting the huddle – it will typically revolve around sharing information people need to know for doing their work.
- How often and for what length of time? Team huddles are typically daily and last from 10–15 minutes. A cross-functional group might meet once or twice a week to review what is happening with a process where responsibility is spread across multiple groups.
- During the meeting, just focus on key information relevant to the participants:
  - Share any important communication.
  - Focus on the exceptions; do not do a report on things that are working effectively in huddle meetings (for example):
    - Any known problems or exceptions for the day?
    - What issues exist relative to the day's work activities?
  - If you are using a Visual Performance feedback board, make it part of the discussion.
  - Validate any decisions that were made during the discussion and note who is taking responsibility for any action commitments.
  - Periodically in these meetings, take a few minutes to express gratitude or appreciation for the work that people are doing. If somebody did something extra, recognize it. A sincere thanks is meaningful communication.
- Do *not* solve problems during the huddle meetings (other than someone committing to an action).

If you do team huddle meetings, I challenge you to run those meetings to get more in touch with reality. That means if problems exist, you have created an environment where people are comfortable discussing them. You don't want to solve problems in a huddle meeting, but pointing out issues that exist is good. *It will require some courage* to do this.

In my field of performance improvement, people like to use the metaphor of lowering the water level so we can see the "rocks" in the stream impeding flow. This sounds like a great idea. But the personal challenge is publicly displaying your problems. We are used to being responsible for taking care of our own problems. *I don't want you to see my rocks because you will think I'm not doing a good job if you see them.* Leaders who are highly effective at improving go the other way, and sometimes you need to be the pioneer leading the way to more effective leadership practices. There is a risk in doing this, and courage is required. But leaders brave enough to operate this way increase trust levels with their team members and, ultimately, their peers.

Gary Peterson from O.C. Tanner said,

> We use "work team huddles," for us this is 3–8 teammates gathered at the start of their workday around their huddle board. The huddle lasts no more than 15 minutes. The purpose of the huddle is to learn from yesterday, and plan for success today. Topics could include Quality, Cost, Time/Delivery, Safety, Appreciation, Short Training on Lean, etc., today's experiment for Improvement Kata, etc. All the data should be displayed on the *huddle information* board for all to see.
>
> About seven years ago, our Team Leaders were running the huddles, and even though team members looked engaged, you never really knew what they were thinking. So, we decided to better connect all team members to the huddle by giving them ownership of one of the pieces of data. On the huddle board, next to each chart, is a picture of the team member who owns it. When the meeting gets to their topic, they step forward and lead the discussion.[16]

I wouldn't start with team members being required to speak in a huddle session, but after a leader has been doing this for 3–6 months, you have created a standard for conducting the meeting. At that point, reaching out to team members and having them more actively participate in leading part of the meeting sounds like a great idea.

Move from a Telling Model to a Coaching Leadership Style

We touched on this subject earlier with "how to ask better questions" and in the Reflection chapter (Chapter 2) with Google's list of ten traits of effective leadership behaviors.

A coach's role is to unlock potential and enable people to execute the fundamentals of their particular discipline more seamlessly. Part of the challenge of becoming a good coach is that you must let go of your "auto-mode" way of operating. It is so easy to tell people how to do something. It takes less time, and you feel like you did a good job because "you told them!" This is something that requires reflection and clarity of purpose. *Are you focused on the short term in your coaching or the long term?*

I was never a "natural" athlete. When I first started playing baseball on a Little League team, I had trouble hitting the ball. I was swinging the bat pretty hard, but most of the time, I missed the pitch. After my first couple of outings, the coach (my best friend's father, Harvey Grant) walked up to me and asked, "Michael, what do you see when you swing at the ball?" I was sort of dumbfounded, not sure how I responded. He then said, "The next time you go up to the plate, I want you to tell me if the ball goes under or over the bat (*if you miss the pitch*)." I said, "Sure coach." In my next at-bat, I missed the first pitch, but noted it went under the bat. On the next pitch I swung again, and I saw the bat hit the ball! I was so excited, I almost forgot to run. But I did make it to first base and was quite proud of myself.

When I was sitting on the bench after that inning, the Coach asked, "So what did you see?" And I explained both swings, the first one under the ball and the second when I hit the pitch. Coach then said, "You just learned the secret to being a good hitter; always let the bat hit the ball; you should be able to see it." I became a much better hitter after that simple coaching session because I watched the ball meet the bat. This is the first coaching experience I remember. It made an impression because I still remember how calm Coach Grant was in his advice, very unlike the father-coaches on other teams. It was easy to listen to him. And I felt, with his guidance, that I'd found the solution on my own.

What does good coaching look like? Well, once again, it depends on the situation.

Coaching Model – Toyota Kata

If you are skill-coaching your team members on how to improve or pursue goals, there is a lot of value in studying the model outlined by Mike Rother. He has done research on how Toyota teaches problem-solving based on scientific thinking. A key part of it is an effective coaching process. *Toyota Kata* describes patterns of thought and action inside Toyota, and Mike provides a practice method for organizations outside of Toyota to operate with these dual targets.

"Kata" are patterns or practice routines. You can apply Kata-type language in your conversations for coaching team members. One of Rother's Starter Kata's for the coach, outlined below, is based on a five-step iteration of learning and improving and follows the framework of five basic daily questions:

1. What is the target condition? (Understanding the direction)
2. What is the current condition now? (Grasping where you are)
   Now reflect ...
   a. What was your last step?
   b. What did you expect to happen?
   c. What actually happened?
   d. What did you learn from the experiment?
3. What obstacles do you think are preventing you from reaching the target condition? Which "one" are you focusing on?
4. What is your next step (plan/do/check/act problem-solving experiment)? What do you expect to happen?
5. When/where can we go to see what we have learned from taking that step?

Once the target condition date is reached (typically two weeks out), establish the next target condition based on what you know then. Begin the next cycle (Figure 4.6).

The model and how to practice are explained in much more detail in *Toyota Kata* and the Toyota Kata Practice Guide.[17]

Coaching Model – TWI

If you are training a new operator or a new associate on how to do a specific task, telling might not be all bad if you include a reason why the steps you are describing are important. A model called Training Within Industry (TWI) follows a telling, showing, and explaining format. This methodology was originally developed within the United States in the 1920s. It was later exported to Japan after WWII, and because people had forgotten about it in the United States, it came back to the country during the Quality Crisis in the automotive industry (1970s). The developers of this methodology believed supervisors needed to have:

1. Knowledge of the work (e.g. how to do things).
2. Knowledge of responsibility (what needs to be done, by when).
3. Skill in improving methods (how can we do this better?).
4. Skill in leading people (why we do things this way).
5. Teaching ability (how to pass along our skills to others).

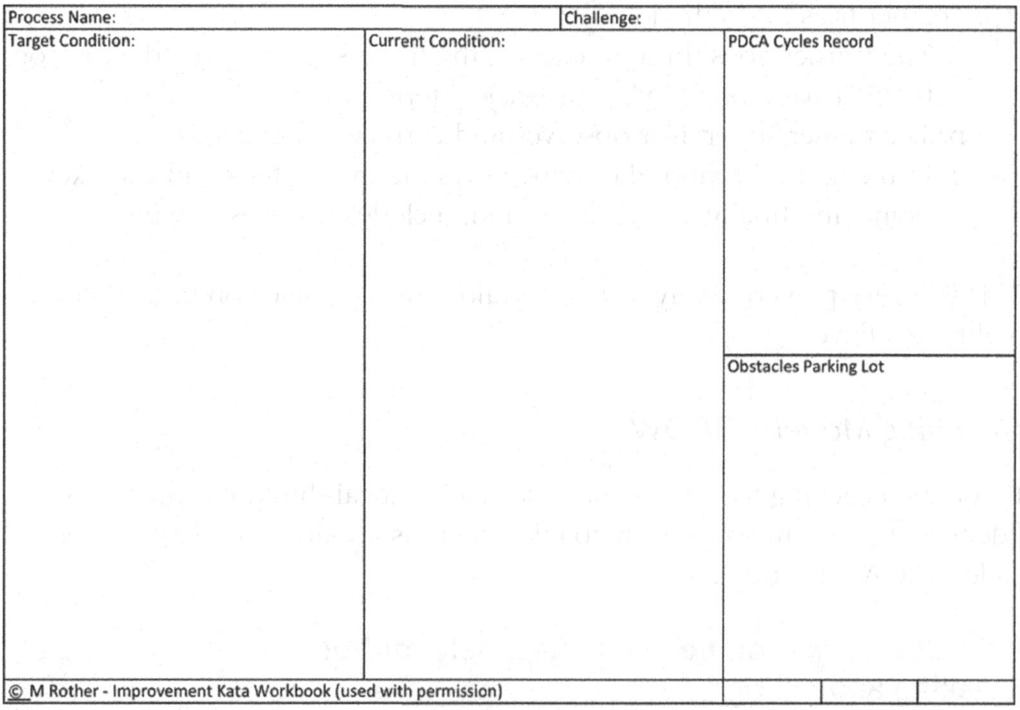

Process Name:		Challenge:	
Target Condition:	Current Condition:	PDCA Cycles Record	
		Obstacles Parking Lot	
© M Rother - Improvement Kata Workbook (used with permission)			

Figure 4.6   Mike Rother Kata worksheet

Source: **https://www.slideshare.net/slideshow/kata-slides-graphics/37399091 (Slide 22). Used with permission from Mike Rother**

The job instruction training is quite specific in its format:

1. Prepare the learner (put them at ease, find out what they already know, get them interested in learning, and put them in the correct position for the task being taught).
2. Present how to do the task to the learner:
   a. Do the job and describe the major steps.
   b. Do the job, state the major steps, and stress each key point.
   c. Do the job, state the major steps, key points, and explain the reason why.

So, the trainer goes through three passes, building the next step into each pass.

3. Learner tries to do the job:
   a. The learner goes through each of the three steps described above on the first pass, only saying the major steps.
   b. The trainer/supervisor observes and corrects if needed.
   c. In the second round, the trainee says the major steps and each key point, and finally, in the last round, includes the reason why.

This is a very powerful way to coach/train new associates on tasks that are highly repetitive.

Coaching Model – GROW

If you are coaching to develop an associate's critical-thinking skills, confidence, or problem-solving capabilities, there is another coaching model called GROW (Figure 4.7).

GROW model can be used effectively for both asking and telling

		Goal	Reality	Options	Wrap-up
Ask	Reflect	It sounds like you'd like to focus on...(?)	So the work is going well overall but you're concerned about...(?)	What I'm hearing is that you see three options...(?)	So you're feeling confident of being able to follow through ...(?)
		These are all clarifying questions...not reflective			
	Ask clarifying questions	What would a successful outcome look like? What specifically should we focus on?	What reasons did they give for the change? How would you account for your success?	What would be involved in pursuing that option?	Exactly when do you think you could complete that task?
	Ask facilitating questions	What do you want? What would you like to get out of this session?	Where are you now? What's working? What's not working?	What are some options for change? What resources might you use?	What are some steps you can take? What obstacles might you face?
	Ask challenging questions	What would be a goal that would represent a breakthrough for you and for the organization?	What prevented you from saying what you thought directly to others?	If you knew the answer, what would it be?	On a scale of 1 to 10, how committed are you to carrying out the actions you have described?
Tell	Assert	Given what you've said, I think we should focus on...(What do you think?)	I see you making assumptions that you may not be aware of...(What's your reaction?)	I see another option that I'd like to put on the table...(What do you think?)	I think you need to do A first, then B, for the following reasons... (Does that make sense?)

Figure 4.7   GROW model questions

1. **Goal:** What is the goal of the coaching session? What is the employee's goal, what is she trying to accomplish? Goals should follow the SMART format:
   a. Specific.
   b. Measurable.
   c. Attainable.
   d. Realistic (could be a stretch).
   e. Timely or time frame.
   Questions could include: What are you seeking to accomplish? Why? What outcome would be ideal? Why? What is the target? What is your target? Why?

2. **Reality:** What is the current reality of your situation? Questions here might focus upon: What is happening right now? Why? What progress have you made so far? Or what is inhibiting your ability to progress? The idea here is to get a handle on the current situation (reality) and the desired future state.

3. **Options:** What needs to be done to reach the goal? Questions here might focus on: What is your first (or next) step? How else might you do this? What might happen if you do that? Who else does this impact? Who should be involved? When was the last time something didn't work the way it was supposed to? What happened when the problem occurred? Were any actions taken; if yes, by whom?

4. **Will or way forward:** What are the action plans for the next steps? Questions here might focus upon: What needs to be done right now? How are you going to do it? How will you know if this step is successful? Who else does this impact?

There is plenty of material available on the web to further describe these models and others. If you use some of the simple assessments described elsewhere in this material (e.g. Trust Index – W.L. Gore, Clarity of Purpose exercise), it can help you become a more effective coach. Good coaches provide feedback, ask questions, show empathy (relate to the person), listen, point out both strengths and weaknesses, and guide the coachee (where appropriate) to arrive at their own solution or next step and provide some structure for the process. I typically like to ask a person being coached at the very start of our process: Why do they want to be coached? What are they hoping will be different as a result?

In a conversation with Joyce Russell, President of Adecco's North American business unit, she said,

> Talent scarcity is our biggest problem. The people we want to hire are working somewhere else. We need to create a value proposition to attract good people. Our talent determines who we are. It's what makes us special. It defines everything about us: our culture, our energy, our competitive advantage, and ultimately, our position in the marketplace. My job is to grow and develop the next generation of leaders at Adecco; that's how I leave a legacy, by the people I hire and mentor. *The true meaning of leadership is to plant trees, under whose shade you do not expect to sit.*

Leader's Standard Work

This is a powerful and very under-appreciated tool. People understand the need for standard work practices for those who do repetitive tasks like assembling a product or working in a call center. If you've never done that type of work, there is a tendency to think it is mind-numbing and rigidly repetitive. But even in those environments, there is a surprising amount of variation and need to figure out how to do things.

When you move to tasks like R&D, project management, or being a leader, people tend to believe "standard work practices" reduce flexibility and are overly restrictive. Sort of like wearing handcuffs (Figure 4.8).

And it is true if Leader's Standard Work is defined as a "Task To Do" list, it can be very restrictive and feel like you are trapped or it's a complete waste of time. But if done well, the exact opposite can happen (Figure 4.9).

"Through discipline comes freedom" is a quote attributed to Aristotle. Standard work done in the right way makes it much easier to focus on further improvements and in seeing/observing variations.

There is a best practice for doing work, for leading, for coaching, for doing an R&D experiment or performing surgery, and for a myriad of other activities. Leader's Standard Work (LSW) should focus on what is the best possible way to do a job.

Figure 4.8   Trapped

Figure 4.9   Breaking free

When creating LSW for your role, give some consideration to the following:

■ What are the most important things you need to do on a daily, weekly, or monthly basis to develop people and improve processes and business performance? Identify those activities. Your job should not be primarily focused on hitting the current target. If that is what you are currently doing, then you are not doing a very good job of developing your people and improving the stability of the processes used by your team to do work. You must break out of the short-term (today) focus and on the longer term. In *The 7 Habits of Highly Effective People*, Steven Covey said, "too often leaders get sidetracked by short-term problems and things that are not important. They miss the important but not urgent task."[18] *LSW should help you better address what is important but not urgent.*

■ You may want to create some type of checklist; do not make it a laundry list. Routine tasks you do daily can simply be called other work and meetings. Focus on the things that are not urgent but are important. What do you want to be held accountable to do? You might even break your checklist into multiple times per day, daily, weekly, etc., buckets. Items on your list might include:
  – Team huddle meeting (daily).
  – Check staffing levels vs. work plan (daily).
  – Gemba Walk (might be multiple times per day, each with a different purpose/focus).
  – Morning leadership meeting (these are important to list if multiple people are trying to do LSW so you can lock-in the same time of day).
  – Review performance against plan (probably happens several times a day).
  – Coaching (might be a daily or weekly activity).
  – Problem-solving, continuous improvement work (daily).
  – Express gratitude for work being done (daily/weekly … more spontaneously than a fixed time slot), etc.

■ We talked about "purpose" in Chapter 3. What is the purpose of doing LSW? It typically should focus on three "Ps":
  – Improving people.
  – Improving process.
  – Improving business performance.

If your LSW activities don't focus on doing those three things, then your LSW actions are not fully unlocking your abilities to lead. Figuring out how to touch those three things as a leader on a daily, weekly, and monthly basis requires reflection and practice.

When people first do LSW, it typically looks like a task list (meetings, Gemba Walks at certain times of day, reports that must be written, etc.). That is not what LSW should be.

We discuss the importance of standard work in the *How to Do a Gemba Walk*. It also fits with "Visual Leadership" because if there is a standard, it should be observable. When I visit a company, I typically observe a leader as they do their standard work during the day or during a Gemba Walk. I can observe how a project manager sets up a project or how a surgeon preps her team just prior to surgery. We want a best practice for doing these activities and then getting as many people as possible to do and then improve the best practice.

Put a plan together for holding yourself accountable to do this, and, if possible, get periodic coaching on how to do it better.

Challenge Your Assumptions

Prof. Robert Carraway at the University of Virginia writes about avoiding "analysis paralysis." He suggests we challenge our assumptions. We tend to think our assumptions are facts. We need ways to challenge those beliefs. In a blog, he wrote,

> One way to identify these (assumptions) is to ask oneself, 'If I knew right now that despite my intuition, this course of action was doomed to fail, what would be the most likely reasons?' These most critical assumptions then become the drivers of analysis, targeted at validating – or refuting – the critical assumptions on which the success of the course of action will ultimately hinge.[19]

We should commit at the very front end that if the analysis refutes your assumption or your belief, you're willing to consider your assumption/belief is incorrect. Stating this at the front end makes one more open-minded to the results of the analysis. It's a great exercise for intellectual humility.

If you gain a deeper understanding of the "real" purpose behind why work is being done, it will automatically make you more of a cross-functional process thinker and help you break out of a silo mentality.

Additionally, if you are getting more clarity about developing people and improving cross-functional process activities, you will consequently enhance the performance of your organization.

Better Leader Learning Cycle

Did you ever go to a party, and it felt like everyone simply wanted to talk about how great they were, and no one asked you a question? We have been to a few like that.

It can be similar at work. At work "talkers" typically tell people what to do and how to do it. They don't explain "why" this is important or what the purpose is, nor do they seek to learn the associate's understanding of the purpose. This often continues to happen even after a new employee has gained experience and has ideas. Questions seldom get asked to learn about the new associates' point of view. *Make certain that you are not a talker.* Learn to listen more effectively and develop the critical-thinking skills and accountability for taking action with your team members.

In a presentation, Paul O'Neill (former Secretary of the Treasury and past Chairman of Alcoa) said, "You show respect for people if your employees can answer 'yes' to three questions"[20]:

1. Am I treated with dignity and respect by everyone I work with (regardless of my position, ethnicity, etc.)?
2. Do I have the knowledge, skills, and tools (support) to do my job?
3. Am I recognized (appreciated) and thanked for my contributions?

Leaders who practice behaving in a way consistent with this list show a holistic respect for people.

In the *Great Places to Work Report 2016*: "The arc of the moral universe is long, but it bends towards justice." Martin Luther King, Jr. once said,

> Human beings have moved from a long era of enslaving each other to an industrial age where workers were free but often abused, to the adoption of workplace rights in the 19th and 20th centuries. In the 21st century, there has been growing recognition that organizations, employees, and society all benefit when workplace cultures do much more than meet basic labor standards such as safe conditions and wage and hour laws.[21]

What did you learn from this chapter? What are your thoughts on the following about building more effective relationships?

- Adopting Leader Standard Work practices to improve the way you lead.
- Learning to ask better questions.
- Become a better or more effective coach.
- Using a Toyota Kata model to coach effective improvement practices.
- Expressing gratitude and providing more positive feedback to your associates.

- **Potential Actions**: What are the two actions you can execute in the next 30 days to build better relationships in your area of responsibility, your team, and your organization?
- **Practice**: What is your practice plan for those actions? When will you start (date/time)? How will you hold yourself accountable to do it?
- **Evaluate**: How might you validate that any new actions taken are more positive than negative in terms of impacting your team, your department, and with your peers?
- **Next Steps**: What do you plan to do next because of what you just learned? Nurture and celebrate the diversity within your work team. Create a safe space for people to grow, and you will be rewarded as a leader.

Outstanding leaders are ordinary people who move beyond what average leaders accomplish because of the choices they make and the actions they take. We all have the ability to do this; the choice is yours to make.

Notes

1. *Harry Chapin: When in Doubt, Do Something*, 2021, Greenwich Entertainment Company.
2. Original quote attributed to Carl W. Bueher in the 1971 collection titled *Richard Evans' Quote Book*.
3. https://www.leansparx.com/single-post/2017/12/20/Support-Others-and-Create-New-Sparks.
4. http://www.gallup.com/businessjournal/182321/employees-lot-managers.aspx.
5. http://www.gallup.com/opinion/chairman/212045/world-broken-workplace.aspx?g_source=EMPLOYEE_ENGAGEMENT&g_medium=topic&g_campaign=tiles.

6. New Trader U blog – Steve Burns interview with Steve Jobs 2007.
7. https://blog.gembaacademy.com/2019/04/12/deliberate-reflection-create -improvement/.
8. https://mike-robbins.com/work/.
9. *Bring Your Whole Self to Work*, Hay House Publishing May 1, 2018.
10. I did not fact-check this quote, but I love the idea it expresses. Not trying to slam the Swiss. ;o).
11. https://www.gatesnotes.com/2018-Annual-Letter?WT.mc_id=02_13_2018_02 _AnnualLetter2018_DO-COM_&WT.tsrc=DOCOM Bill & Melinda Gates foundation February 13, 2018.
12. https://jobs.netflix.com/culture.
13. http://www.thehighvelocityedge.com/book Steven Spear; McGraw-Hill Mar 3, 2010.
14. https://www.youtube.com/watch?v=tAb5tgVrkg8.
15. https://www.linkedin.com/pulse/leadership-round-table-mark-preston/?trackin-gId=Yj%2Fxl2hLTpu4lMxa%2BA38Jw%3D%3D.
16. Gary Peterson, Workshop at AME Annual Conference, October, 2023.
17. http://wwwpersonal.umich.edu/~mrother/Handbook/Appendix.pdf.
18. Steven Covey, 7 *Habits of Effective People*, Simon & Schuster, 1989.
19. https://ideas.darden.virginia.edu/2015/10/the-most-important-and-least-asked -question-in-business/.
20. https://hbswk.hbs.edu/archive/paul-o-neill-values-into-action.
21. http://learn.greatplacetowork.com/rs/520-AOO-982/images/GPTW-2016-Global -Report-Final.pdf.

Write Your Thoughts on Building Relationships Here

Chapter 5

Visual Leadership

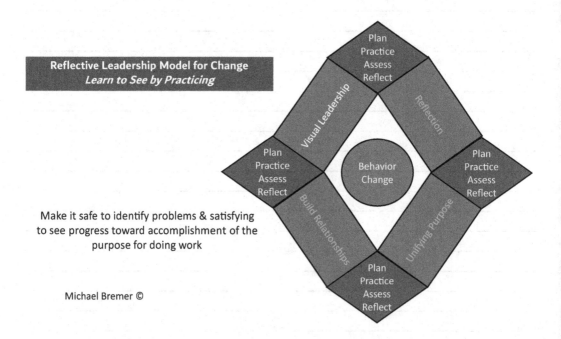

Reflective Leadership Model for Change
Learn to See by Practicing

Plan Practice Assess Reflect

Visual Leadership

Reflection

Plan Practice Assess Reflect

Behavior Change

Plan Practice Assess Reflect

Build Relationships

Unifying Purpose

Plan Practice Assess Reflect

Make it safe to identify problems & satisfying to see progress toward accomplishment of the purpose for doing work

Michael Bremer ©

Why Is It Important to Make Information Publicly Visual?

The primary purpose of visual performance metrics isn't to make sure things are working well in your department. It's intended to create a thinking environment which makes it easier for multiple departments, teams, and groups to work together. It is relatively easy to come up with performance metrics for your team, but what about metrics that help "us" to work more effectively

DOI: 10.4324/9781003495284-5

together? What about metrics that cause us to think? Good visual report-ing practices create an "information democracy,"[1] a phrase expressed by Filippo Passerini, CIO at P&G. They eliminate filters that obscure knowledge between layers of management and between departments. They help to cre-ate an environment that is much more robust and open, making it easier to be in touch with the "actual reality."

When metrics are publicly displayed, they are a reminder to everyone of what is important. They also serve as a communication mechanism. When you post data, both the people who supply information, services, or physical products and the people your team serves can see the data. If anyone dis-agrees with the data, a conversation can take place to clarify or correct the information. Once there is general agreement about the information being posted, good visuals nurture improved performance between groups.

Many years ago, Rummler and Brache wrote a book about managing the white space. They were referencing the white space on the organizational chart between all the departments. Their point was that when things go wrong, the handoffs between cross-functional groups often are part of the cause. This is very similar to Deming's "most problems result from the pro-cesses we use for doing work."[2]

Publicly posting key metrics seems to accomplish several positive things:

- If other teams/departments disagree with the metrics, they will come forward and let you know. It promotes more open and honest discus-sions between siloed work groups.
- Visual metrics provide focus. However, just making them visual doesn't work any magic. Making decisions and taking action give credibility to the numbers.

Visually posting information makes it easier for cross-functional groups to work together. It also makes it easier for people to point out the "ugly babies" if leadership makes it safe for people to say it. *Even if your organiza-tion isn't doing anything along these lines, there is no reason you can't do this with your team and interact with your peers to make certain they agree with the data being posted.*

As you make your first visuals, give some thought to the purpose of the visual. What problem(s) are you trying to solve by having a visual. Are you simply seeking to understand the current status? If yes, what does that mean? Is it activity counts? Is it making problems easier to see? What will this visual do to help you, your team, and your interactions with your peers

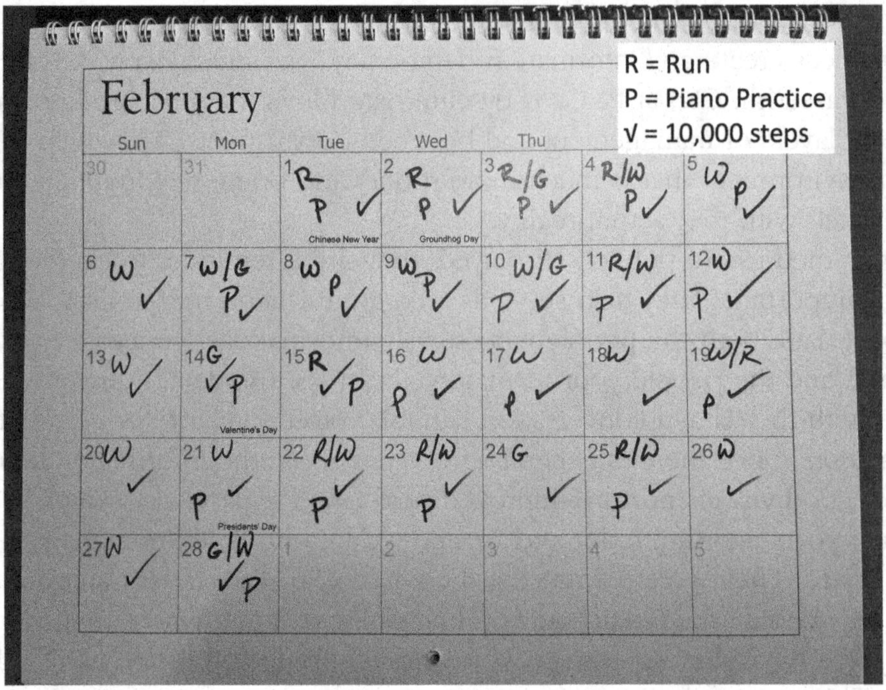

Figure 5.1   Calendar

work more effectively and make better decisions more quickly? Then do an experiment – try it! Create a visual with your team members and see if it comes close to meeting your purpose for the visual.

One of my decisions for the year 2022 was to learn how to play the piano. I also wanted to make sure I walked at least 12,000 steps every day and to revive my running to five times per week. My very simple accountability practice was to note on a physical calendar that hangs in my home office if I accomplished the task. Here is a calendar shot from February 2022 (R = Run, W = Walk, P = Piano, $\sqrt{}$ = 12,000 steps, G = Golf) (Figure 5.1).

I clearly need to work on my "R" running target. I was doing okay in learning to play the piano. Find some simple way to hold yourself accountable for what you wish to change.

One Leader's Experience: Engineering Department – Request for Proposal Process (RFP) for Custom Designs

This example comes from a company on the west side of Chicago that manufactures fractional horsepower motors. The Design Engineering Department had a large backlog of projects. Jordan, a young engineer in that group,

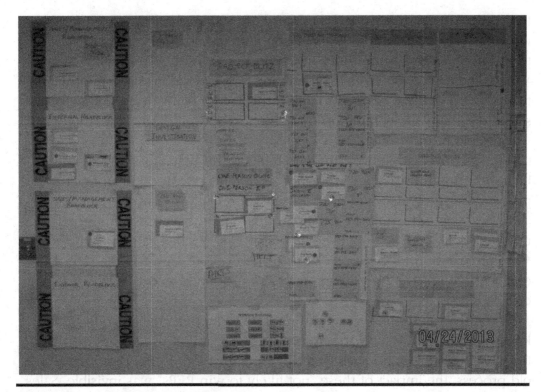

Figure 5.2   Design engineering board

decided to create a picture of their capacity for doing work. His original intent was to use this tool to manage their capacity. I'm changing the actual numbers here for confidentiality. They had five design engineers in the group, and after doing some analysis, he realized that, on average, each engineer could work on five proposals at any point in time. So, the department's capacity for doing work was 25 RFPs or 5 RFPs for a single engineer. This picture is purposely not readable (Figure 5.2).

Initially, the Design Engineering Group used the visual board to level their workloads. There was some variation in the complexity of the proposals. So, if John was working on his five but fell behind schedule because two of them were more complex than normal, Mary might have been working on a mix of proposals that were easier to do. Looking at the visual board, she could see John was falling behind and offer to help. No one had to ask Mary to do this. She could easily see the need by looking at the board.

Under the old process, there was also not much active collaboration between Purchasing, Sales, and Operations. Work activities got passed from one functional group to the next without understanding the capacity to do the work. After Jordan posted the board, they started having twice-weekly meetings with leaders from Engineering, Purchasing, Sales, and Operations to review the status of the RFPs in queue as part of their bi-weekly Gemba Walk standard work activities. Purchasing and Operations used information they learned in these reviews to adjust their work plans because they could see things that were coming up over the next several days. Sales, though, was the big breakthrough.

These conversations very much changed the behaviors and dynamics between the groups. They were not terrible to one another under the old process, but they also had no genuine appreciation of how their work activities were affecting the other groups. There was now an agreed-upon capacity for the department of 25 RFPs. If sales came in with RFP #26, a discussion took place. "Which one do you want to take out of the queue if we insert #26?" This conversation opened people's eyes to some other issues.

When the salespeople were getting RFPs from potential customers, they knew they were not likely to win several of them. But if a salesperson won the RFP, it was great because they got a commission. In the past, under the old process, the impact of these "not likely to win RFPs" was invisible. Sales knew there was an impact on Engineering, but they didn't see it.

Once they began to participate in these twice-a-week meetings to review the RFPs in progress, it became publicly obvious how many RFPs were winners and how many were not; they were winning about 33% of the time and losing the rest. A 67% loss rate is not good! Intellectually, people knew they did not win most of the proposals. But now that they were looking at it on an information board, posted for everyone to see, with information that was not disputed, their (sales) view of the world changed.

No one told Sales they had to do this, but over the next several months after they started using the visual board, Sales began to do a better job of vetting the proposals. Quite a few of them were unlikely winners (e.g. the customer was primarily seeking a comparative bid). What the salespeople did was to stop taking proposals they were unlikely to win. They also did a better job of getting the information Design Engineering needed to draft the proposal. Over the next 12 months, the RFP win rate climbed from ~ 33% to over 60%.

This story demonstrates the power of good visuals. They really can help people work more effectively together in a very natural way without anyone forcing the issue.

Visual Leadership vs. Visual Management – Things to Think About

Part of the reason for talking about publicly posting key metrics is to drive behavior change inside your team, department, or organization. The story in the last section shows the power of good visuals. They really can help people work more effectively together naturally, with no one forcing the issue.

Many companies that started using visuals have more of a management viewpoint with their visuals rather than a leadership perspective. Visual management seems to primarily focus on the results being accomplished now. There is power in doing that, but it doesn't go far enough; behavior change is needed to excel. *Visual leadership* is more focused on behavior changes. What behaviors, practiced more effectively in the future, will improve our performance? Behaviors might include being more open about problems, doing a better job of problem-solving, finding problems more quickly, and increasing people's confidence and capability to take on more responsibilities, etc.

One of my first experiences with something close to visual reporting happened in the 1980s. I had been given the responsibility to create a company-wide productivity improvement effort for Beatrice Foods. We were very decentralized, with 440 profit centers or businesses (as mentioned earlier) across the globe. If some bloke in the corporate office (like me) wanted to do something, the profit centers were not *required* to do it.

I held that job for about four years in total. Two years into it, I was curious if anything was being accomplished. So, I created a visual board (summarized by Division) that was kept in my office and a report (by profit center) with a family of metrics (this was prior to the publishing of the Kaplan/Norton balanced scorecards article). The results were updated quarterly. The family of metrics selected included Return on Net Assets, Operating Margins, Revenue Growth, and Employee Turnover. All this information was easily available from our corporate database.

We called our improvement effort *Strive for Excellence*, and participation was voluntary. This was early in my learning about effective improvement practices. About 40% of the profit centers were taking part two years into the program. When I compared a three-year track record of the profit centers that were taking part versus those that were not, I was surprised! Before creating this report, I really did not know if we were making a "real" difference, and I was still uncertain what effective improvement practices looked like from a holistic perspective. But lo and behold, the 40% of the profit centers that were taking part were showing more improvement in those metrics than those who were not.

Prior to creating this report, I had not shared quantitative results with anyone. After creating the "Family of Metrics for our Productivity Improvement Process," I was requested to share the report's results quarterly with the senior level executive team. Over the next three months, participation levels in *Strive for Excellence* climbed from 40% to 75%. The increase in participation primarily happened due to more transparency with the current participants and the senior executives' increased awareness.

Hidden Problems Don't Get Solved

Most leaders try to address problems when they see them. And once you figure out something that works, people like to keep doing it. In many ways, this is what successful organizations, teams, and departments do; as we become successful, our existing approaches get reinforced, and we want to keep doing things the same way. Unfortunately, this can become a strength carried to excess, which then turns into a weakness. So, finding new ways to experiment and continuing to reflect on what you are trying to accomplish and how you are going about it is important.

What happens when you don't see it? There is a long history of very successful leaders who had blind spots that ultimately doomed their companies. Or they reached a level in the hierarchy where they no longer led effectively. Problems remained hidden, perhaps due to the leader's assumptions, lack of experience or a changing external environment, or their personal "fears." These leaders were not receptive to the fact that there were problems they could not see and possibly refused to listen when others tried to shine a light. Because they were unaware of their blind spots, they assumed the problems did not or could not exist. Silicon Graphics, Compaq, IBM prior to Lou Gerstner, Arthur Andersen, Nokia, Toshiba, MySpace, Motorola, Blackberry, DEC, The Gap, and a long list of companies either ceased to exist or missed out on major business opportunities due to blind spots and misguided beliefs.

And since employee engagement surveys from Gallup, Wyatt, and others typically show that 70% of the people who work for a company are not actively engaged in their work, then it's reasonable to believe that *if the CEOs of some organizations have blind spots, then it's likely this is also true for other leaders inside those businesses.*

Visual Leadership, practiced effectively, helps us to shine a bright light on problems and then to engage people in solving the issues. In many ways, Visual Leadership provides the sustenance and energy to keep working the other three Key Foundations. Your goal as a leader is to create an environment where people are not afraid to say, "Here is a problem!" and then figure out how to address the issue(s). You need to uncover and reveal what is not seen.

Visuals Shouldn't Be Complicated

Visual does not necessarily equal numeric. There are a variety of visuals that can provide useful information. They used one of my favorites in the HR Department at Autoliv in their Logan, Utah facility. Figure 5.3 shows their format. *The data shown in this visual is not real. I made it up.*

They were using this graphic to track ergonomic risk. It was posted near the exit door from the HR Department that entered the factory, and it showed where they had identified ergonomic risk relative to the way people were doing work. The purpose of the visual was to refresh people's minds when they walked into the plant about places where they had issues last year and the current year. The behavior they were trying to reinforce was when they went into the factory, they wanted to always be thinking about

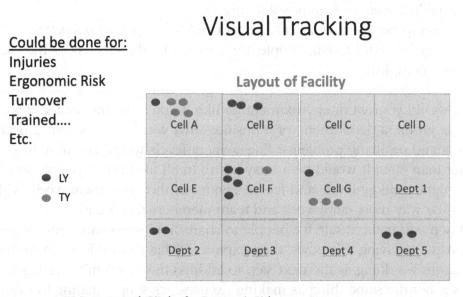

Figure 5.3   Simple visual

safety. And even if they were going out on the floor for another purpose, they could still glance at work being done in cells where there were issues to see if the new procedures were being followed or if changes implemented last year had been sustained. It took less than a minute to update this board, and everyone I observed going to the floor seemed to practice looking to see where issues existed before they walked out of the door. People could use this simple graphic for many things: Infection rates in a hospital, employee turnover in a production cell or department, etc.

Be in Touch with Reality – Drive Away Fear

Recently, I heard Isao Yoshino, a retired Toyota leader, speak on this subject. Mr. Yoshino was a mentor for John Shook many years ago at NUMMI, which was a joint venture between Toyota and General Motors. Mr. Yoshino said, "My aim was to develop John by giving him a challenging mission or target and support him while he figured out how to reach the target. As I was developing John, I knew I was developing myself as well."[3]

Mr. Yoshino also shared several leadership lessons:

- Share bad news first; no problem is a problem.
- Go to the Gemba yourself vs. people bringing you reports to learn what is really happening.
- Establish seemingly impossible targets.
- Develop people to attain the target (*that is your job as a leader*).
- Never be afraid to fail; people are honored by the challenges they seek to accomplish.

What would your work environment be like if you were the leader of a team, a group, a department, or a business that was full of people constantly finding and resolving problems? Engagement levels would certainly be higher than 30%. It would be a busy place. In all likelihood, people would have higher energy levels and feel in control of their destiny at work, which is not the way most employees and team members feel today.

When you make it safe for people to share "bad news" first, you decrease fear. After receiving this news, then explore, "What did we learn from this?" "What are we doing as the next step to address the problem?" Sharing bad news is not the same thing as making excuses. It's simply getting in touch with the reality of the situation and being willing to openly address key issues. Sometimes those issues will be performance-related to the individual, but typically these issues arise from the processes being used to do work.

Dr. W. Edwards Deming said, "85% of all performance problems inside a business are directly related to the processes being used by the business, only 15% are related to the individual."[4] Later in life, Dr. Deming was using a 95% to 5% ratio. I find that a little hard to believe, but it is totally believable that most problems inside organizations result from the way internal processes operate. Make sure metrics are being used to drive improvement; avoid using them as a weapon. Dr. Deming often said, "We need to drive fear out of the workplace."[5] Many performance measurement systems do exactly the opposite. When management doesn't look at the business process and instead focuses on someone, some (other department), or some outside factor to "blame," it causes people to game the system and point their fingers elsewhere when problems arise.

Think about the word process with a capital "P." For example: If a new employee is having problems doing work, is there a problem with the onboarding process or the employee training process? If parts are not being delivered on time to the manufacturing floor, is it the Purchasing Department's fault, or is there a problem with the way the company is ordering component parts? If manufacturing costs are too high, is that the fault of manufacturing, or does the new product development process typically fail to consider manufacturability early in the design process? *Just changing your perspective to look at issues from a process perspective first rather than trying to find someone to blame can go a long way to eliminating "fear" inside an organization.*

What do you do when you walk up to new parents taking their baby for a stroll in their baby carriage? Virtually everyone says, "My, what a pretty baby!" No matter what the baby looks like. That is not how you want your team to work. Where are your ugly babies? Where are ideas that are not fully baked? How can you give them an opportunity to metamorphose into a butterfly? You want people to shout, "My, oh my, that is one ugly baby!" If people get comfortable doing this, you will not have so many hidden problems. Employees and leaders will be more in touch with reality as it really exists rather than the reality we wish was true (where everything works smoothly most of the time). *PS ... I'm not literally suggesting you tell anyone they have an "ugly baby."*

Billy Ray Taylor, a retired senior executive from Goodyear Tire and Rubber, likes to reference the *"10 Second Rule."* Whenever he looks at a visual board, he believes the observer should know within 10 seconds where the issues lie. The best boards make it easy for observers (even casual observers who do not work in the area) to see where the problems are. Visual Leadership practices should fully support finding your "ugly babies" and doing something about them.

Shifting gears now to what should be made visual. Give some consideration to the following thoughts.

Great Metric Test

First, look at the metrics used by your team, department, or company and compare them to the four circles in the Great Metric Test (Figure 5.4):

1. Does the metric have a valid relationship to the successful accomplishment of the "purpose" for doing this work activity?
2. Is it timely? Is it received too late to act? That does not mean it is a bad metric, but it is certainly useless for real-time action.
3. Do people understand the metric? Is it a credible number that people believe they can positively influence? Do they think the metric is fair?
4. How many metrics are you using? If you have a laundry list of 20 or more items, probably several metrics will conflict with one another. How do people know which one counts, and which one is important? Try to focus on a handful of metrics that drive improvement.

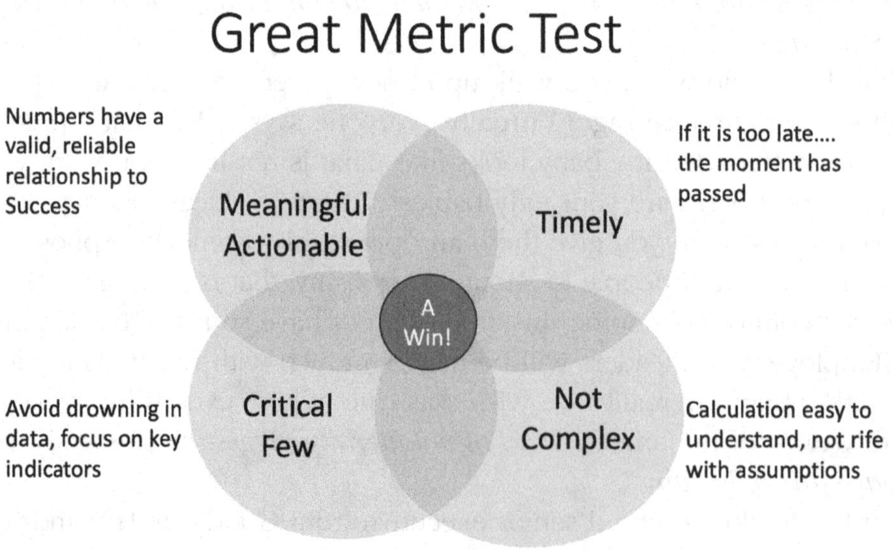

Great Metric Test

Numbers have a valid, reliable relationship to Success

Meaningful Actionable

Timely

If it is too late.... the moment has passed

A Win!

Avoid drowning in data, focus on key indicators

Critical Few

Not Complex

Calculation easy to understand, not rife with assumptions

Figure 5.4   Great metric test

The intersection of those four circles reveals what may be some of the most important metrics to accomplish your purpose – a win!

Traditional vs. True North Metrics

Metrics that most people use are lagging indicators (after the fact), and they are activity counts (how many or how much did we do). They will let you know when you have a problem, but those types of measures rarely proactively drive improvement.

True North Metrics is a phrase often used by companies adopting Toyota-type (or Lean) improvement practices. True North Metrics require a blend of perspectives. Typical categories include (Figure 5.5):

The metric categories shown below are a reasonable list for thinking about what you should measure. The specific metrics for each of those categories would differ for manufacturing, software development, hospital patient care, etc. But the categories are universal. For sure, they don't apply to every situation, but initially, check to see if you have key metrics that measure those attributes as a starting point. You will need to tailor the specific metrics being used for your situation.

If an internal group is serving another department, then that group is your customer. There are many internal customer–supplier relationships

'True North' Metrics

Blend of Perspectives

- Customer
  - Value Proposition....for the right customers
  - Loyalty, Referrals, Revenue Growth, % New Products....
- People (respect)
  - Engagement, Development, Participation in Improvement....
- Quality
  - Process Capability, Defects, First Time Through, Performance, Reliability, Conformance, Through the Eyes of the Customer.....
- Velocity (Time)
  - Delivery, Lead Time, On-Time & Complete, Takt, Cycle....
  - How close can you get to 100% Value Added Time
- Cost ....
  - Productivity = Wealth (how can you capture it)
  - Financial, Risk.....

Figure 5.5   True North Metrics

inside organizations. But typically, performance is measured from the individual silo's perspective of what is good, rather than a cross-functional process perspective, which is ultimately how customers get served.

So, we need to move beyond those narrow perspectives to find metrics that drive the right behaviors and the right results.

Finding Leading Indicators

It's very normal to focus on output metrics and after-the-fact outcome metrics because it's what you see at the end of the day. They are useful in focusing attention on key business results. However, they are not useful in managing upstream business processes to get better results. It takes some detective work to move upstream to find meaningful "leading indicators."

If there is a "magic sauce" in metrics, it is finding relevant leading indicators that can be managed to predictably get the results we want. But they're not always obvious, so we must search for effective leading indicators. One way to do this is whenever someone suggests a metric, ask "what drives the performance of that metric?" Do this several times, plumbing down beneath the surface for a deeper understanding. This simple technique often reveals powerful leading indicators that can foster early corrective action.

A methodical process for the causal metrics search can look like the graphic in Figure 5.6. The four steps are:

1. Define the results to be monitored.
2. Identify the process steps that directly lead to the output results.
3. Identify any inputs that contribute directly to the output results.
4. Test for causality by evaluating the cause-and-effect relationships between the upstream/causal metrics and downstream results.

Over time, leading indicators should be tested to see if they truly correlate to a win. Going one step deeper is to understand causality. Just because there is a correlation between "A" and "B" does not mean there is causality between A & B. In an ideal world, there would always be plentiful data available to test causality with statistical analysis tools. However, the real world is often not that convenient, so hands-on examination and process experience are usually reasonable substitutes.

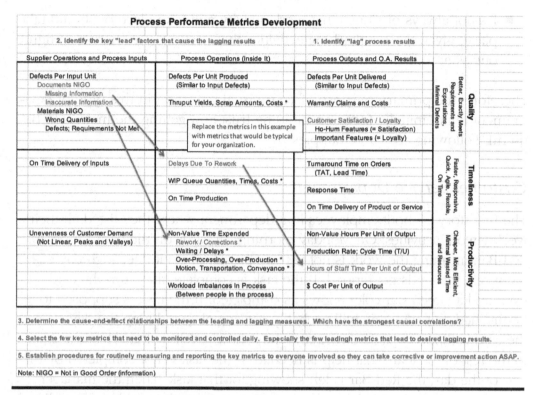

Figure 5.6   Four steps to find leading indicators

Examples of Visual Leadership in an Office

One thing to keep in mind if you try doing Visual Leadership in an office environment is that at first, people will probably be nervous about the concept. Performance metrics for office-related activities, be it engineering or accounting, are usually not measured in the same way as physical processes (manufacturing) or highly repeatable transaction processes (call centers). Also, people in an office environment are typically multitasking; they are most likely not doing highly repetitive activities. So, they are not used to being measured. And their first thought is likely to be something like, "Oh my gosh, I'm not exactly sure why we are doing this?" So, they get concerned about performance metrics and people observing their work activities.

■ People being observed feel threatened by the activity. They are not used to being measured publicly and wonder what will happen as a result (it couldn't possibly be good).

■ It's not always obvious what work activity will take place or should take place, given the multitasking activities that occur. So, it's harder to figure out what to measure.

Therefore, the primary purpose of your visual reporting in this type of environment might initially focus on building trust and getting the metrics to be useful. Following are two examples:

Customer Service Department

The first story is about employees in a department where they listed all their weekly work activities. On Monday morning, all cards on the board show a Red Bar, meaning the work is not yet completed, nor is it due. Each day of the week, the cards get flipped. If the task is complete, they turn the card over, and a Green Bar color shows as completed. If the task is supposed to be finished, but someone will probably not complete it during the day, then the card is flipped upside down, and they show a Yellow Bar. This is a signal to the rest of the team that help is needed. If the task is not completed by the end of the day, the Red Bar is left showing for that task. Then the manager knows there is an issue and can take steps to resolve the problem. They listed the key steps for the work task on one side of the card (Figure 5.7).

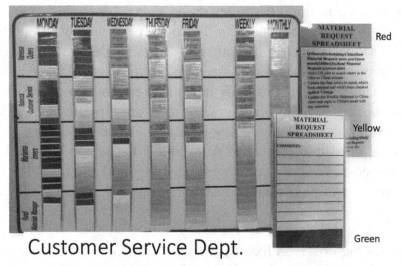

Figure 5.7   Visual tools – customer service

Employees use the cards to see who is on schedule and who has an excessive workload. Their normal condition is that some days one employee will be busier than the others, and one employee may have a lighter workday. When they look at the board, it is easy to see who is running behind and who can offer to step in and assist. This works great as work ebbs and flows between the employees. If one person consistently has problems keeping up, then there is a deeper analysis to understand what is happening from a process perspective. The team felt this helped them work more effectively together; they were the people who designed and used it.

Problems with Visuals in an Information Technology Group

This second story comes from the IT department in a Fortune 100 corporation.

> We developed a tool to enable IT services to be assessed against enterprise readiness standards. This initiative started out with the goal of enabling teams to perform self-assessments and take corrective actions where necessary, rather than being audited by IT. The results were publicly posted. We hoped teams would be motivated to increase their scores. Rollout of this tool did not go as well as we'd hoped.
>
> Some application development groups were reluctant to participate, fearing that a poor score would reflect negatively on their performance. Typically, these were teams with legacy services with a lot of known issues, outside of the team's control. They were building shared services for use by other internal business units across the company. They were concerned that the self-assessment scores would be visible to their customers, and a low score would give their customers a reason not to use their service. Although these objectors were in the minority, they were vocal.
>
> Use of the tool varied significantly. Some teams tried hard to get a 100% score and were paranoid that less than a perfect score would get perceived as a failure. Others performed an assessment, got a low score, and did nothing more; they felt the task was to complete the assessment, not to act on the results. Despite regular evangelizing of the tool, adoption dropped off quickly after launch.

There was also a perception issue. An initial misunderstanding that the tool would be used to "audit your services" rather than "help you make your service enterprise-ready," created adoption challenges. The assessment tool also needed ongoing development and support, but there weren't resources available to do that. The conclusion the assessment tool team came to was that the tool was too complex to maintain and operate, and the lack of a clear value proposition made further investment difficult to justify against other initiatives.

On the positive side, however, this concept materialized in another form. As part of a cloud migration program, the cloud engineering team developed a dashboard for managing all the cloud services owned by an employee. Besides listing the services, the dashboard audits the services for cloud and security policy compliance and flags violations. Many times, remediation can be done with a single button click. It's a great example of how to make issues visible and actionable. Collected metrics were reported to executive stakeholders and they looked at them from a process perspective vs. finding someone to blame. It created a closed loop that drove the right behaviors and organizational benefits. It addressed the "fear of being audited" by putting the responsibility for compliance right in front of the user.

This is an interesting story as when we first spoke, they were very excited about the tool and just rolling it out to the department. There are several lessons to be learned:

- Tools by themselves are not enough, no matter how well-intentioned.
- The purpose for using the tool must be clear from the start. It helps to have the people who will use the tool take part in defining its purpose.
- The outcome they wanted to accomplish was vague. They initially wanted to drive an improvement in enterprise readiness across all IT services, which is a good thing from a leadership perspective, but the reason for doing so was unclear.
- You must go the last mile. It's not enough to put the tool in the user's hands. Make it easy for them to identify the problem and take action to remediate the issue.

The initial experiment was ultimately a success! One of the IT teams created a more holistic tool that did several things. Relative to compliance, it not only checked compliance but it also suggested remediation steps when a compliance issue existed. So, there was more ownership by employees with the new tool, it was easier to use, and policy compliance improved. I apologize for not being able to show a picture of the visual, but the organization was not comfortable releasing it.

Patience to Get It Meaningful

With some patience, discipline, and tenacity, you can make progress with Visual Leadership. Your task will be much easier (not easy) if leaders behave in the role of a servant leader who is there to see that barriers get removed. It's also best, as these examples demonstrate, to clean-up your own backyard first before trying to improve other departments or work groups. Over time, credibility will be built, people in the office will develop skills and confidence, and significant change can take place.

Remember, on this journey, you are using the visuals to learn how to more effectively execute the other three foundations: Reflections, clarity of purpose, and elevating your team and your peers. Your first attempts at doing this may not work because your first boards are likely to be too complicated, take too long to update, and may not provide the insights you hoped to gain. That's okay; this is a journey.

Visual Leadership is the inchpin providing substance and ongoing energy to sustain these practices and to hold yourself and your team accountable. I strongly urge you to try it.

Actions to Practice – Visual Leadership

Exercise to Create Your First Visual

1. Involve the people on your team and share an initial purpose for creating a visual reporting tool.
2. Consider something important to your success, but difficult to see or perhaps subject to different interpretations inside your organization, your team, or with your peers.

3. What would be the purpose of the visual feedback for that item (keep refining the purpose, why is this important)?
4. Create a rough outline of potential elements on the board (visual).
    a. What information should be present (time, activity, quality, productivity, improvement, progress toward a target etc.)?
    b. Consider creating a visual that shows your capacity for doing work (remember the example of the engineering department) – work capacity is typically invisible in an office environment.
        i. Is there a backlog? Do people continue to give you work even though you cannot handle it?
        ii. Is there a rework/redo/additional requirements type problem?
5. What decisions might your visuals drive? (Note: this is important to think about.)
    a. Would it surface things that need to be improved? Might it provide timely insights?
    b. Might it drive more collaboration?
    c. Probe to better understand what drives performance?
    d. Search for leading vs. lagging indicators.
6. How would you keep the info up to date (it needs to take less than 15 minutes daily, ideally less than 10 minutes)?
7. Share the draft with the rest of the team, perhaps also with suppliers and customers, for the work by your team.
8. Change the visual based on feedback received and then try it for a week, a month, or a quarter.
9. At the end of your initial experimentation period, evaluate what is happening with the visual compared to your original purpose.
10. Continue to use and evolve.

Going Deeper – Focus on Key Process Drivers

First, realize this is totally an experiment at the start. You are not likely to get the metrics or visuals right on your first or even your second pass. Give some thought to "WHY" you should publicly post metrics. Write it down since this is a key factor in evaluating the success of your experiment.

Reread the Design Engineering story previously discussed. There is a lot to be learned from that example.

- The various functional groups involved in the RFP process talked to one another about the process regularly (twice per week).
- A picture is indeed worth 1,000 words.

Metrics also need more focus on "process drivers" if they are going to influence early corrective action. Consider the model:

Inputs ⇒ Process ⇒ Outputs

Most performance metrics focus on process outputs, not the actual process itself. For example, "Late Deliveries" is an output metric for the Delivery Process. An item is delivered late or on time (assuming early is not a factor). But the final delivery time results from many upstream in-process actions, some of which can cause considerable delivery variability if they are not monitored and controlled. Those key controls are sometimes called "process drivers" because they literally drive the process in a particular direction – hopefully, on the right road – so both the drivers and the roadmap better be clear to everyone involved.

It's important to identify the drivers with causal relationships to output results. A helpful tool for that is a cause-and-effect (fishbone) diagram. Start with a question like "What are the possible causes of Late Deliveries?" Identify a few categories for the main "bones" in the diagram and work down each of them to identify as many causes as possible. Asking "Why?" five times is also a good way to get farther down each of the bones in the diagram.

At different times, some drivers will affect a process more than others that currently have only a minor impact on the outputs. Consider the 80/20 Rule: 20% of the process performance drivers probably have the major impact at any point in time.

Output metrics will not go away. They are the easiest to compare on a period-to-period basis. Give thought ahead of time to the key performance drivers that affect the critical output metrics. If something slips out of alignment, then simply dig down into the next level of detail to provide more useful information to address the key process issues.

Realize that process metrics are an iterative focus on what is important to improve given the current situation. An effective measurement system should be dynamic enough to rotate different drivers onto the radar screen to monitor process health. The drivers may be looked at for a three-month, six-month, or twelve-month period. After the process gets stabilized, the process driver metrics on the radar screen may change as new issues emerge.

Use a Balanced Set of Metrics

Norton and Kaplan wrote about the power of a Balanced Scorecard in an HBR article many years ago.[6] As was noted earlier in this chapter, in the world of "Lean" performance improvement activities, the idea of "True North" metrics gets stressed. Select a handful of metrics, and as my friend George Koenigsaecker (former CEO at HON Industries and an early Lean advocate who helped to create the Danaher Operating System) says, if you improve them every year, "good things happen." The list below has a few ideas of potential metrics to use. Typically, you should focus on:

- Safety, in an environment where physical work is being done (also applies to offices ... seen several people over the years injured walking downstairs).
- Quality, for obvious reasons.
- Delivery/lead time/cycle time (bottleneck).
- Productivity improvement (Art Byrne, former CEO of Wiremold, said, "Productivity = Wealth").[7]
- Capacity utilization (this is especially important to consider in an administrative environment because the capacity to do work is typically invisible).
- Human development (again, I would hope for obvious reasons, this is more than hours trained).
- A future performance target (What does better look like 12 months from now?).

Ideally, your organization has some type of Policy Deployment or Hoshin Planning process. Your key metrics should link to the key strategies/ objectives for the business, where practical and meaningful. George Koenigsaecker talks about the power of "true north" metrics in his book *Leading the Lean Transformation*.[8]

Figure 5.8 is a spreadsheet we have used to determine if a balanced set of metrics are being used, whether they are they leading or lagging (most lag), and whether they are they linked to the organization's key strategies.

Performance improvement metrics should start with the customer in mind. No actual surprise here. The surprising thing is how often this does not happen. Just using a few metrics focused on meaningful customer issues can drive behavioral change and reallocation of resources. *Note: these various worksheets are loaded on the michaelbremer.net website under the Books tab.*

Metric Effectiveness Worksheet

Functional Responsibility (Summary or Mission):

Department/Group:

Contact for this Worksheet:

Metric Classification(s)

Metrics	Customer Sat.		Timeliness		Quality		Productivity		Cycle Time		Employee Development		Cost		Risk		GMT	Cust.
	Lead	Lag	Lead	Lag	Lead	Lag	Lead	Lag	Lead	Lag	Lead	Lag	Lead	Lag	Lead	Lag	Pass/ Fail	H/N/I

Medium Relationship ●

High relationship √

		# Pass	# Fail	# Help	# Neutral	# Inhibit	# Leading

Metric Analysis

5	Apply the **Great Metric Test** - Must meet at least 3 for a **pass**, if less it **fails** (GMT)	#
6	Check with your internal customers to see if your metrics **help**, or are **neutral** or **inhibit** services to **their customers**	%
7	How many metrics are listed?	
8	Decide if your metrics are appropriately linked, a balance of views and a few early warning indicators	

Instructions

1. Describe the primary responsibility or mission for your work team, group, or dep't

2. Enter key metrics currently utilized

3. Indicate if the metric has a **high** or **medium** relationship to the various classifications. Leave blank if low or none.

4. Determine if the metrics are leading or lagging indicators by catergory - some may belong to serveral classifications (change names if needed)

CUMBERLAND

Figure 5.8   Metrics effectiveness worksheet

Source: Created by the author. Used with permission of the Cumberland Group

Standard Work Makes It Easier to See What Is Happening

We talked about Leader's Standard Work in the Build Relationships chapter (Chapter 1). Standard work practices also make it easier to see if the right thing is being done at the right time for associates.

Steps to create standard work:

- Involve the people who do the work in creating the standard.
- Determine the handful of things that make a difference if done consistently.
- Group the things you do into families or groups, so that:
  - a consistent way of doing this work is defined,
  - people are trained, and
  - discipline is monitored.
- Make work more visible so everyone knows the "right" things are done at the right time.

The details needed for Standard Work (SW) vary based on the type of job being done. The SW for tasks that are repeated multiple times per hour will be much more structured than something that is less repetitive on an hourly or daily basis. It also varies based on a person's experience.

A senior operator can probably use a high-level checklist. She does not need detailed instructions. A new person needs more guidance than an experienced associate. The SW applies to briefing a surgical team, to launching a project, and to software development and work being done on a factory floor. In all cases, we want people to use the best practice as we currently understand it.

List of Things That Could Be Visual

This list is simply to get you started thinking about what could be visual. It is not intended to be all-inclusive.

Safety:
- Displays of hazardous materials.
- Safety lines (green or yellow walkways).
- Dangerous points (high voltage, chemicals, etc.).
- Markers to show pallet or container heights and storage locations.
- Markers to show the risk of injury (fingers, hands, etc.).
- Emergency exits and evacuation plans.
- Safety control board.

- Safety cross.
- Number of days accident-/incident-free.
- Number of safety ideas implemented.
- Safety inspection results.

Prevention of Defects:
- Poka-yoke (mistake proofing) markings (alignment, error-proofing guides, etc.).
- Yellow containers for questionable or unknown parts.
- Red containers for defective parts.
- Visual aids, pictures, and charts to identify Standard Work steps.
- Quality control board.
  - Graphs with defects, good parts, first pass yield, etc.
  - Warranty claims.
  - Delivery issues.
  - Quality improvement action plans.
  - Number of days with no defects.
  - Pictures of good/bad parts.
- Quarantine areas and persons in-charge.
- Scrap and/or yield rates.
- Hospital-induced infections.
- Maintenance, preventative maintenance – done on time, done correctly.

Employee Development:
- Skills matrix (Figure 5.9).
- Number of ideas submitted and implemented.
- Attrition or turnover rates.
- Time to competency.
- Successful practice of targeted new behaviors (see how to measure in Reflection chapter [Chapter 2]).
- Engagement (usually a total business metric/visual).
- Internal promotions (more of a total plant or company visual).
- Value added per associate or employee (sales minus purchased materials divided by total headcount) – *also more of a total plant/company metric.*
- Safety, Quality, Delivery, Improvement, Productivity – Shift 1 & 2 example (Figure 5.10).
- Quantity produced.

Example: Skill Matrix

Process	Cut	Mill	Sand	Frame	Assembly	Paint
Mike						
Bill						
Cindy						
Matt						
Norm						

Figure 5.9   Example: Skills matrix

- Quantity produced on-time.
- Quality of parts, information produced.
- Schedule attainment.
- Productivity (typically some type of output/input ratio), but figuring out what goes into the denominator and the numerator is a challenge. If you can get a handle on the capacity for doing work (see the Design Engineering Group story previously discussed), it can make it easier to rough out a productivity metric. At AME, we use the value added per associate to get a rough idea of productivity for a company as a whole (sale).
- Capacity utilization (how much could we do, how much did we do).
- Changeover times (project to project, machine changeovers, operating room changeover from patient to patient, or within an ER room on bed utilization).
- Lead time from start to finish (new product development, software development, manufacturing from order to delivery, etc.).
- Cycle time (typically a process limited by the step that takes the longest to do – also known as the bottleneck).

While it may feel comforting to say that all metrics need to be maximized, it is simply not the reality. Trade-offs exist. Customer service levels vs.

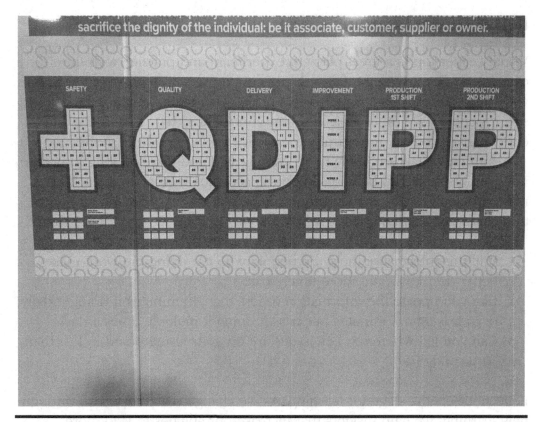

Figure 5.10   Safety/quality/delivery/improvement/production/productivity

inventory levels; on-time delivery vs. resources to make it happen; responding to customer needs to customize offerings vs. process capabilities or resources to accomplish the task. Measurement is relatively simple, but selecting the right metrics is difficult. Depending on the level of the organization and the ultimate connections between processes, the definitions of output vs. process can change. Just look at the earlier example of late deliveries. In the delivery process, this is a process output with a related set of process drivers sitting underneath. From a customer satisfaction perspective, on-time deliveries may itself be a process performance driver. Therefore, it is important for the leadership team to agree on what is most important to improve. This has a profound impact on the proper or improper allocation of resources focused on improvement.

Better Leader Learning Cycle

Visual Leadership is a great place to experiment and practice with the changes you are trying to affect. Do this in partnership with your team and with your peers in other departments or on other teams. Do not get defensive if people have other ideas. Probe to understand why they see it differently. It gives you an opportunity to practice asking good questions.

Your first visuals are likely to be complicated and take too long to update. They are also likely to be focused on the activities your team does. That is not a problem, as this is a learning exercise. As you experiment, give some thought to these three rules:

1. Will this information help the people who are doing the work make better decisions on a more timely basis?
2. Can you update the information in less than 10 minutes if doing it daily or in less than 3 minutes per pass if doing it multiple times a day?
3. Can you tell where the key issues are on your visual board in less than 10 seconds?

Reflection sets the stage for learning. As we change our perspective through new learning, we gain insights that increase our abilities to lead more effectively.

- **Potential Actions**: What are two actions you can take to experiment with Visual Leadership? What is important in your area of responsibility that should be done in a more effective way? Look at the above three rules and create your first visual.
- **Practice**: What is your practice plan for the visuals? How will you use them (hourly, daily, weekly)? How will you gain buy-in from your team?
- **Evaluate**: How will you validate that the information on the board is reasonably accurate? How might you validate that any new actions taken are more positive than negative in terms of affecting your team, your department, and your peers?
- **Next Steps**: What do you plan to do next because of what you just learned? Visual Leadership provides feedback on what you learned from your reflections. It lets you know how effectively the purpose is being accomplished. Visual Leadership practices should help to drive cooperation and collaboration with your team members and peers in other departments.

The most important choice you make is what you choose to make important.

Michael Neill

Notes

1. Brian Wilson blog post WSJ, https://blogs.wsj.com/cio/2012/05/07/pg-cio-filippo-passerini-discusses-big-data/.
2. https://deming.org/extraordinary-results/.
3. Katie Anderson, *Learning to Lead, Leading to Learn*, Integrand Press, 2020.
4. https://deming.org/extraordinary-results/.
5. Ibid.
6. Robert Kaplan and David Norton, *The Balanced Scorecard – Measures that Drive Performance*, Harvard Business Review, February 1992.
7. "CEO's Role in Lean Transformation," Mark Graban Podcast 158, https://www.leanblog.org/2012/09/podcast-158-art-byrne-the-lean-turnaround/.
8. George Koenigsaecker, *Leading the Lean Enterprise*, Taylor & Francis Group, 2009.

Write Your Thoughts on Visual Leadership Here

Chapter 6

How Do You Change a Habit?

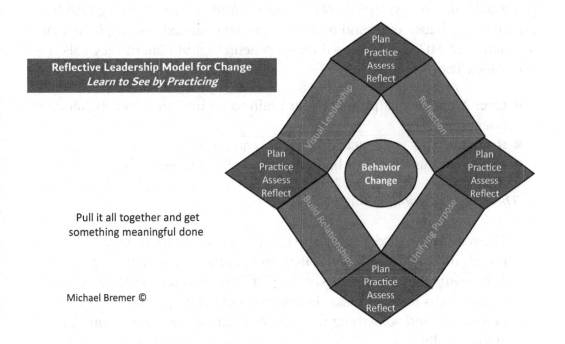

Reflective Leadership Model for Change
Learn to See by Practicing

Pull it all together and get
something meaningful done

Michael Bremer ©

Changing Your Behaviors

Now let's look at bringing it all together. Your learning is going to go a lot faster if you can find a few other leaders who also desire to make some changes. You can form a small support group that might meet on a weekly or bi-weekly basis to share your plans, your progress, and your challenges. Having a peer group will help foster a greater sense of accountability, and

DOI: 10.4324/9781003495284-6

you can learn from one another. Absent that resource, you can still do this on your own. It will require some tenacity and discipline. But you can do it. If nothing else, update me on your progress. Many of the forms referenced in this book are available under the "Books" header at my website https://michaelbremer.net.

Pulling It Together

Letting Go of Old Habits

I am asking the reader to look at the way you lead and consider if there are some changes you would like to make. Are there new habits or routines that you would like to develop? In *The Power of Habit*, Charles Duhigg (2012) explains why habits exist and how they can be changed.[1] He explains that researchers at MIT described a three-step neurological pattern they called "The Habit Loop" (Figure 6.1).

- **Cue:** A trigger that prompts your brain to go into an automatic mode of operating.
- **Routine:** This is the actual behavior and the related action(s) taken.
- **Reward:** This tells your brain that the habit is worth doing.

In *The Power of Habit*, Duhigg wrote,

> The reason the discovery of the habit loop is so important, it reveals a basic truth: *When a new habit emerges, the brain stops fully taking part in decision-making.* It stops working so hard and diverts focus to other tasks. Unless you deliberately fight a habit, unless you find something new, the old routine pattern will unfold automatically.[2]

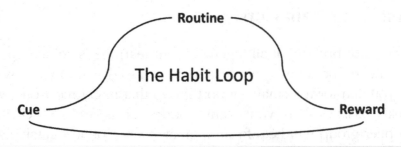

Figure 6.1   The habit loop

It's going to take some effort to change your old habits. The more immediate the reward, the easier it will be to develop the habit. Since immediate rewards are not likely to happen in changing the way you lead, this book describes the considerable importance of Visual Leadership in Chapter 5 as a technique to help inculcate new habits/behaviors.

In summary, the steps you are going to take to form this new habit include:

- **Potential Actions**: Based on the results of your reflections, what actions do you plan to execute in the next 30 days to better define the purpose for your leadership, your team, and your organization?
- **Practice**: What is your practice plan for those actions? When will you start (date/time)? How will you hold yourself accountable to do it?
- **Evaluate**: How might you validate that any new actions taken are more positive than negative to affect your team, your department, and with your peers?
- **Next Steps**: What do you plan to do next because of what you just learned?

Now that you have a basic understanding of how habits form, let's focus on developing a few new habits. Let's start with a short story about Martha. She is a real person (not her real name). Except for saying Martha read this book (she did read an early draft), everything below is largely true. I simply condensed the actual events to fit the space in this chapter. Martha still works for the same organization today, serving at the Senior VP level, and has global responsibilities.

Example: One Leader's Experience Changing the Habit Loop

Martha Johnson leads a small plant with 90 employees. It includes a production environment, a small engineering group, and a couple one-person support departments (HR, Finance, Purchasing). When she looked at the way her team operated, she felt a few changes were needed because of problems in the operation, notes from her boss, and customer feedback. The plant runs reasonably well, and they almost always hit their numbers, but the day-to-day environment feels a tad hectic. There is an excessive amount of firefighting taking place to get things done. She wondered if she needed to make some changes as a leader.

Martha is also reading this book. After reading the Reflections chapter (Chapter 2), she wondered about their current way of operating and some of her behaviors as a leader. She spoke to her leader, Carol Smith and several of her associates. She then decided she wanted to experiment with the material and start with the Reflections chapter.

When she looked at her day-to-day operations, the team members did quite well on their repetitive responsibilities. The work associates producing their products knew their jobs and performed timely and quality work. However, completing their task/action activities always seemed to be problematic. While some of this was surely normal, why did they often restart a new task? Sometimes people seemed to procrastinate getting started, leading to a scramble at the end to complete the work or resulting in work that got done was not what she expected to see.

Reflection

Martha decided to get some initial input from others as she began reflecting on how she and her team handled their task/action activities. She spoke with all seven of her direct team members. She decided to use the Keep/Stop/Start exercise described in the Reflection chapter Actions to Practice – Keep/Stop/Start with her team members. She asked:

■ "What should I keep doing as a leader that is helpful to this team and to your success when we have new tasks/actions we need to accomplish?"
■ "What, if anything, should I stop doing that might be making it more difficult to get work done?"
■ "What, if anything, should I start doing that will help us all grow and be more successful?"

She gave everyone three 2 x 2 Post-its and said they were not obligated to write anything, but she would appreciate their assistance. The feedback was anonymous, and she requested they put the Post-its on her office window before going home that night. In the morning, Martha had 20 pieces of feedback. Several of them are:

- **Keep:** Trusting team members to do the work.
- **Keep:** Opportunities to learn when doing work activities outside of day-to-day responsibilities.
- **Stop:** Two people felt pressured to step up and say, "Yes! I'll take responsibility for a new task."
- **Stop:** Several staff members felt Martha was not always clear with what she tasked them to do when initially giving them an assignment.
- **Stop:** Quite a few people were feeling a bit overwhelmed by the amount of "special projects" that needed to get done.
- **Start:** Three team members wanted more coaching from Martha on how to proceed with their task.
- **Start:** Several staff members felt the priorities were not always clear on what action was most important to take.
- **Start:** Making certain staff members underst and "why" this is an important thing to do and how the actions might impact people outside of the department.

Martha reflected on trying to get a better understanding of how she delegated these new work activities. Looking at the "Habit Model" she noted:

- **Cue:** We have a new problem or opportunity that needs to be addressed.
- **Routine:** In our Friday meetings, I give a brief description of the problem and ask if anyone can take responsibility for it. Sometimes, I'll assign it to an individual.
- **Reward:** I feel like I've taken the appropriate first step in getting a resolution, and I move on to the next thing on my to-do list. I feel a little bit of relief at that moment.
- **Performance:** We are struggling to get things done. There is too much going back and restarting, and sometimes work that I think is happening isn't, so we must scramble to complete the task. It's energy-draining for everyone.

Martha discussed the feedback from the staff and her reflections with the team in their Friday staff meeting. There were several things people felt were somewhat invisible:

- Priority of the project relative to other actions and day-to-day responsibilities.
- In some instances, Martha's expectations for the assignment were unclear.
- Perhaps the workloads were not evenly or appropriately distributed among staff.
- If Martha had done a little more coaching earlier in the process, team members were less likely to wander down the wrong (or less effective) pathway.

After the meeting with her team, Martha went back to reflecting on the discussion and again looked at the habit model to see how she might want to change the way she led:

- **Cue (did not change):** We have a new problem or opportunity that needs to be addressed.
- **New Routine:** In our Friday meetings, I'll still briefly describe the problem and ask if anyone can take responsibility or assign it to the appropriate individual. But we will stop and discuss the action: Does the person have time to accept this task? Is this a "quick to-do," or will it take longer? Why? When can they start the task (does not need an immediate response)?
- **Reward:** If we can develop a better way to keep track of these items, they are more likely to be accomplished, and it should decrease the time we spend on them. We can also develop stronger relationships with staff as we review their progress.
- **Performance:** We don't have much data on our success rate or rework load looking backward, but we can devise a way to see how well we do on this.

Purpose

Martha gave some thought to the purpose of her old way of operating. It was not written, but in essence, it was *"to make certain someone was covering all open actions."* She used her feedback to move to a more "coaching style" of leadership and to develop the critical-thinking skills of her team members more rapidly. She also wanted to decrease the time it took for everyone to complete these actions. She wrote: *"My purpose as a leader for dealing with new actions is to use task/action items to develop the critical-thinking skills of staff members and to eliminate false starts."*

Martha wanted to ensure team members had a deeper understanding of the projects so they could experiment and innovate when developing solutions. Martha felt too much rework resulted from insufficient communication with people outside her department, who were impacted by the team's work. She had also experienced a few problems with some of her new leaders needing to understand the scope of their work activity; they were only working on a piece of the problem.

Martha shared the purpose draft with her team, and they suggested a few changes to the purpose statement:

> My role as a leader needs to focus on creating an environment where team members can develop their critical thinking skills and take action to resolve the problems facing our customers. I need to encourage team members' to experiment and develop innovative solutions to the more complex issues they address. I also need to provide regular feedback on how they might accomplish their work activities and grow their capabilities more effectively.

Build Relationships

She focused on uplifting the members of her team. The relationship building started with Martha requesting and discussing the feedback she received from her team members. Some of the newer team members were impressed with her humility and willingness to discuss changing how she led. In the past, Martha assumed things were going well unless a team member approached her with a problem. Several team members thought it was their responsibility to figure these things out, and they didn't want to bother Martha with their problems.

To accomplish the new purpose, a few changes were made. In the Friday staff meetings, people were expected to share the following:

- If a task/action was delayed in getting started.
- If they needed help from another team member or Martha.
- If they wanted to review some of their current challenges/obstacles.

Otherwise, Martha looked more closely at the details of each leader's task/actions when they did one-on-one reviews during the week. In her one-on-one sessions, part of Martha's Leader's Standard Work was to review and coach her leaders on their progress. She did this primarily by asking questions.

Visual Leadership

There were two types of Visual Reports implemented to drive these behavior changes. The first one was a Board for the Friday meeting. It contained each leader's name, their Top 3 task/priority commitments/date to start the action, and targeted date for completion. This was done with Post-it notes on a Whiteboard so they could change the format over the next several months. Team leads stepped into the conference room before the Friday meeting and placed a green check by task/actions on track, yellow where they would ask a question in the Friday meeting, and red where they were experiencing problems.

Martha also wanted to get feedback on her coaching behaviors. She used another tool we discussed in the Reflection chapter (Chapter 2) – Measure Desired Behavior's Level of Effectiveness. Every other week for the first two months, Martha asked her team members to score her behaviors.

- Asking a question instead of offering a solution.
- Mentoring/coaching team members to grow their capabilities.

On a 7-point scale, with "7" being great and "1" being awful, Martha scored a 4 on questions and a 3 on coaching as her starting point. In Week 3, she scored 5 on questions and the same result on coaching, a 3. At the end of the first six weeks, she received a 5 on questions from every team member, except for one person who gave her a 6. Her average on mentoring/coaching moved to a 4.

Over time, Martha became so good with these two behaviors that she was regularly getting 6s and 7s. At that point, she stopped measuring these two items and instead focused on a new behavior she wanted to practice. An example of Martha's New habit development template is in Figure 6.2.

Put Together Your Practice Plan Using the Four Foundations

A habit is a formula the mind automatically follows: You need to change the formula and create a new habit loop. What behaviors do you seek to change? How will you validate that this is a good idea, and how will you keep yourself accountable? Use the actions to practice recaps at the end of each chapter to stimulate your thinking and inspire your actions. Please refer to the referenced chapter as you work through each of the four foundations (Figure 6.3).

Learn to See the Invisible				
New habit development template				
Name: Martha		Date: XX/XX/XX		

Reflection

Give a brief description of the problem and ask if anyone can take responsibility or assign it to the appropriate individual. Have a discussion about the action – Does the person have time to accept this task? Do they understand it? Is this a 'quick todo' or will it take more time? Why? When can they get started on the task (does not need an immediate response)?

Purpose

Role as a leader needs to create an environment where team members can develop their critical thinking skills and take action to resolve the problems facing our customers. Team members are encouraged to experiment and develop innovative solutions to the more complex problems we address. Need to provide regular feedback on how team members might accomplish their work activities and grow their capabilities more effectively.

Build Relationships

The most important change I want to make here is do less telling and more asking questions. I'll work on this when coaching associates. Develop a few simple visuals to improve communications, accountability and our performance.

Visual Leadership

Create a board for the Friday meeting with each leader's name, their Top 3 Task/Priority commitments/Date to start the action and Targeted Date for completion.

Post on my office window a graphic with two questions & a rating scale of 7 to 1:
1. How am I doing with asking good questions when you have a problem?
2. How am I doing relative to mentoring/coaching associate abilities to do a better job?

Figure 6.2 Martha's new habit development worksheet

Reflection

Reflection sets the stage for learning. As we change our perspectives through new learning, we gain new insights that increase our ability to lead more effectively. While I strongly urge you to focus on changes you need to make as a leader first, you can also use this model to focus on changes for your team, department, or organization. Get some early input from a trusted/ respected person. I encourage you to converse with at least one other

Learn to See the Invisible						
New habit development template						
Name:				Date:		
Reflection						
Purpose						
Build Relationships						
Visual Leadership						
You can download an Excel version of this form at: www.MichaelBremer.net						

Figure 6.3 New habit worksheet

person and share your thinking as you do your analysis; as noted, it will be easier with a support group.

Reflect upon new routines/behaviors you may wish to adopt. Don't just run with the first behavior that comes to mind. Consider a few behaviors that might be more appropriate moving forward. What is your current reality?

Some of the questions here and in the previous chapters might be useful to stimulate your thinking, for example:

1. What actions must you take to create an environment where team members feel safe pointing out issues and problems?
2. What actions might you take to foster more critical-thinking skills in all team members and further engage your team members in solving problems?
   a. Problem identification/definition.
   b. Gathering meaningful information.
   c. Analyzing and evaluating alternative solutions.
   d. Reviewing/challenging assumptions.
   e. Communicating in a holistic and effective way.
3. What other behavioral changes are people in your network suggesting for your consideration?
4. Consider using the simple assessment in the Actions to Practice section of the Purpose chapter (Chapter 3) to start a dialogue about "purpose" for leadership behaviors you wish to change or for your work team.

Once you have settled on something you wish to change (don't focus on too many; pick one or two at the start), use the "Habit Loop model" to further probe and gain a deeper understanding of your current behavior:

- What is the cue – the trigger that causes you to exhibit that behavior?
- What is the routine – what is the habit/behavior you would like to change?
- What is the reward – how does that habit make you feel, or what do you think you are accomplishing by behaving that way?
- How does that behavior impact others?

You must apply this same thought process to the new behavior(s) you wish to practice. But first, let's finish walking through the other three foundational steps. Your responses will evolve as you gain new insights.

Define a Unifying Purpose

A clearly defined purpose or reason why you desire to behave in this new way will be helpful in driving/guiding your behavioral change. Clarity of purpose for the changes you seek to make in your role as a leader or for the work done by your team or department can help to prioritize which behaviors/actions are most important to take.

The change you anticipate making is an experiment that may or may not yield the new result(s) you desire to achieve.

Let's develop a purpose for the new behavior.

- Why do you think this is a crucial habit/behavior to change?
- What do you hope to accomplish with this change? Can you set a specific goal that will move you toward fulfilling the purpose of the change?
- What is the new habit/behavior you would like to develop?
- What is your reward for this new behavior – how might it make you feel, or what might you accomplish by behaving that way?
- How do you anticipate that behavior/action impacting others?
- How might you visually measure your progress?

Realize you might miss the target on your first pass. That is why we stress the importance of practice and getting feedback on your progress.

Can you use the questions shared in the Unifying Purpose chapter (Chapter 3) to get started?

- Ask, "Why? We are doing this at least 3 times – why is this important?"
- Take a walk and see how your customer uses your outputs (observe).
- Talk to your direct customer's customer.
- Take the metrics you use and share them with your customers. Learn how your measures impact the work they do.

Build Relationships

This book is about leadership and getting better at doing it. Whatever behavioral changes you explore should be linked to uplifting the people you interact with. If you plan to change, you should NOT do it in secret. Who will you engage in helping you make this change and giving you feedback? Who will be affected/touched by the intended behavior change?

- Can you get someone to coach you on your progress periodically? Your boss, a trusted peer, or even your entire team can provide help. You might also form a group with others who are also seeking to practice learning to see the invisible. It could run like a book club, and you might meet (via Zoom) on a weekly basis to share your progress. Over the first four or five weeks, simply work one chapter each week of this book.

■ These changes require some vulnerability from the leader intending to make the change. Let the people affected by this change know you are trying to make it. I'd suggest keeping this "openness" limited at the outset. People who should know might include your boss, your team, a trusted resource, and possibly someone at home like your spouse. I do not suggest broadcasting the intended behavior change to a bunch of your peers or other team leaders. Get a track record first. Let people come to you and ask what is happening as they observe your visuals.

Make sure that you are Not a Talker

Learn to listen more effectively and develop critical-thinking skills and accountability for taking action with your team members.

Remember the three questions we referenced from Paul O'Neill in the Build Relationships chapter (Chapter 4). "You show respect for people *if your employees* can answer "yes" to three questions:[3]

1. Am I treated with dignity and respect by everyone I work with (regardless of my position, ethnicity, etc.)?
2. Do I have the knowledge, skills, and tools (support) to do my job?
3. Am I recognized (appreciated) and thanked for my contributions?

Leaders who practice behaving in a way consistent with this list show a holistic respect for people.

Visual Leadership

This is important to do! Visual Leadership provides valuable feedback on the usefulness of what you learned from your reflections and lets you know how effectively the purpose is being accomplished. Visual Leadership is a great place to experiment and practice with the changes you are trying to affect. Do this in partnership with your team and with your peers in other departments or on other teams. Refrain from getting defensive if people have different ideas. Probe to understand why they see it differently. It allows you to practice asking good questions.

Your first visuals will likely be complicated and take too long to update. They are also likely to focus on you or your team's activities. That is fine, as this is a learning exercise. As you experiment, give some thought to these three rules:

1. Will this help you become a better and more effective leader? When focused on your team, will this information help the people doing the work make better decisions on a more timely basis?
2. Can you update the information in less than 10 minutes if doing it daily or in less than 3 minutes per pass if doing it multiple times daily?
3. Can you tell where the key issues are on your visual board in less than 10 seconds?

Create Your First Visual

Remember the simple calendar I use for changes I am trying to make (running, learning the piano, 12,000 steps …). Your visuals should not be complex. They should sit where other people can see them. Find a simple way to measure your progress and get periodic feedback.

We covered a way to do this in the Visual Leadership chapter (Chapter 5) as the first action to practice in that Foundation. This is somewhat modified to focus specifically on leadership behavioral change.

1. How might you measure your progress on a behavioral change you want to make as a leader? In the example story in this chapter, Martha just did a simple metric posting on her office window. I still use a calendar in my home office to hold myself accountable for changes I'm trying to make. It's updated daily. However you do this, it should take little time to update, and it must be motivational. I'm not happy if there is an interruption in my daily goal of walking 12,000 steps. Silly as it may sound, I like to put a checkmark on my calendar. Do something for yourself that works.
2. Share your purpose for creating a visual reporting tool with your team and anyone else who happens to walk by who happens to have an interest.
3. Depending on the complexity and frequency of what you are measuring, consider creating a rough outline of potential elements on the board (visual).
   - What information should be present (a specific behavior, time, activity, quality, productivity, improvement, progress toward a target …)?
4. What behaviors or decisions might this visual drive? (Note: this is important to think about).
5. How would you keep the info up to date (ideally less than 10 minutes)?

6. Allow your team members and other interested parties to make suggestions, but keep whatever you do meaningful to you.
7. Try it for a week, a month, or a quarter. At the end of your initial experimentation period, evaluate what is happening with the visual vs. your original purpose and expectations.
8. Continue to use and evolve.

You can download the New Habit Worksheet, Figure 6.3, from my website http://michaelbremer.net under the "Books" tab.

Thanks for Reading and Close

I appreciate your tenacity and hope the information will be useful to you in the next few steps of your life's journey. The goal of this book was to share a few stories/experiences and explain why it is so important for most of us to become better leaders. Hopefully you have already started with some experimentation and practice with the Actions to Practice shared at the end of each chapter.

I genuinely hope you embark on this journey. Experiment, learn, grow, and share with others. The world needs better leadership. Please provide it.

Notes

1. Charles Duhigg, *The Power of Habit: Why We Do What We Do in Life and Business*, Random House, February 2012.
2. *The Power of Habit*, Charles Duhigg, p20, Random House Trade Paperbacks, 2012.
3. https://hbswk.hbs.edu/archive/paul-o-neill-values-into-action 11/04/2002.

Index

Page numbers in italics indicate figure respectively

Printed in the United States
by Baker & Taylor Publisher Services